John Mathew

Eaglehawk and Crow

A study of the Australian aborigines including an inquiry into their origin and a

survey of Australian languages

John Mathew

Eaglehawk and Crow
A study of the Australian aborigines including an inquiry into their origin and a survey of Australian languages

ISBN/EAN: 9783337319656

Printed in Europe, USA, Canada, Australia, Japan

Cover: Foto ©berggeist007 / pixelio.de

More available books at **www.hansebooks.com**

EAGLEHAWK AND CROW

A STUDY OF

THE AUSTRALIAN ABORIGINES

INCLUDING

AN INQUIRY INTO THEIR ORIGIN

AND

A SURVEY OF AUSTRALIAN LANGUAGES

BY

JOHN MATHEW, M.A., B.D.

LONDON
DAVID NUTT, 270-271, STRAND
MELBOURNE
MELVILLE, MULLEN AND SLADE
1899

DEDICATED, BY PERMISSION, TO

THE

ROYAL SOCIETY OF NEW SOUTH WALES

AS A SLIGHT MARK OF APPRECIATION OF

ITS EFFORTS TO PROMOTE

RESEARCH IN SCIENTIFIC SUBJECTS

PECULIARLY AUSTRALIAN

PREFACE

In 1889 I contributed a competitive essay upon the Australian Aborigines to the Royal Society of New South Wales, for which I was awarded the Society's medal and a prize. The present work is based upon that essay, and I desire at the outset to express my warm thanks to the Society for courteously permitting me to make free use of its contents.

But while following the lines of the former work, continued investigation and access to fresh materials have enabled me to amend, modify, elaborate and add much that is new.

As a warrant for venturing into the field of Australian anthropology, I may explain that when a youth I was engaged in station life in the Burnett District, Queensland, in which neighbourhood for a period of seven consecutive years I was in intimate touch with the Kabi tribe. As the fruit of that intimacy I wrote an account of the tribe (containing a grammatical sketch and vocabulary) which is incorporated in the late Mr. E. M. Curr's large work on the Australian Race. During the past ten years I have extended my studies to the aboriginal tribes as a whole.

Compelled by the logic of facts, I have had to take up a new position on various important points upon which at first I had accepted the views of others who had preceded me in writing on the aborigines, especially those of my friend Mr. Curr.

Mr. Eyre's theory (endorsed by Mr. Curr and generally holding the ground), that the first settlement was in the north-

west, and that the distribution of population was effected by
the original stream of people crossing to the south of Australia
in three broad separate bands, I have found untenable. The
distribution of language proves that settlement was first in the
north-east, for there the lines of language converge.

In my paper in the Proceedings of the Royal Society of
New South Wales 1889, I demonstrated as I had never seen
done before, that the language of the extinct Tasmanians was
the substratum of Australian languages, leading to the conclusion that the Tasmanians were the first occupants of Australia,
and settling, I hope, a question which had previously been in
doubt, viz., the relation of the Tasmanians to the Australians.
Further research confirms the view then advanced.

The amalgamation of two races I offer as a probable explanation of the existence of two primary exogamous classes
throughout at least the greater part of Australia, and presumably throughout the whole.

My account of Australian Cave Paintings is an expansion of
a paper which appeared in the Journal of the Anthropological
Institute for 1893.

I attach special importance to the linguistic portions of this
work. The classification of Australian languages is new, and
based upon a comprehensive study of them. That Australian
words have a history, and are not mere arbitrary sounds, is, I
trust, clearly proven. I have shown how some of them have
passed from one end of Australia to another, and have traced
their changes. I hope that the original and systematic treatment of the Australian numerals will be acceptable and the
vocabularies helpful to philologists.

Obligations have usually been acknowledged *in loco*. To
Mr. E. M. Curr's work I am specially indebted.

Authorities for the vocabularies are all given. Several correspondents have contributed valuable information as well as
vocabularies, in response to printed queries. I heartily thank

all informants for the help they have rendered. I have digested
the materials and mentioned peculiar facts.   To publish all that
has come to my hands would be too expensive an undertaking ;
what is unpublished I shall preserve, cherishing the hope that it
may see the light at some futnre time.

JOHN MATHEW.

The Manse, Coburg, Victoria.
  December 6, 1898.

CONTENTS

CHAPTER IV

THE MALAY ELEMENT

CHAPTER V

DISTRIBUTION

CHAPTER VI

PHYSICAL CHARACTERS OF THE AUSTRALIANS

CHAPTER VII

DWELLINGS, CLOTHING, IMPLEMENTS, FOOD

CHAPTER VIII

GOVERNMENT, LAWS, INSTITUTIONS

Government, laws, institutions—Aboriginal bondage to tradition—Tribal cohesion—Leadership—System of kinship and matrimonial restrictions—Ganowanian classes—Blood-ties or marks of courtesy—Dr. Fison on the Murdoo legend—Classes not the result of a conscious reformatory effort—Promiscuous intercourse—Polyandry—Exogamy—Stages of social development as marked by marriage—Australian classes, group-marriage—Negatives as names of communities, class-names and totemism

CHAPTER IX

MARRIAGE, MAN-MAKING, MUTILATIONS, BURIAL CUSTOMS

Marriage, man-making, mutilations, burial—Betrothal—Barter—Marriage by capture—By agreement—Love-letters—Mutual avoidance of mother-in-law and son-in-law—Stages of approach to manhood marked—Initiation to manhood, or the Bora—Primary objects of initiation ceremonies—Mutilations—Circumcision—Amputation of finger-joints—The terrible rite—Mourning—Relics carried—Burial—Death ascribed to sorcery—Cutting for the dead—Abstinence

CHAPTER X

ART, CORROBOREES

Art—Corroborees — Message-sticks of Malay introduction — Rock-paintings, where found—Mr. Giles' discovery at Lake Amadeus—Captain Stokes' discovery at Depuch Island—Mr. Norman Taylor's in Cape York Peninsula—Mr. Cunningham's at Clack's Island—Painting at Nardoo Creek, Queensland—Captain Flinders' discovery at Chasm Island—Captain Grey's at Glenelg River, N.-W. Australia—Authorship—DAIBAITAII—Mr. Bradshaw's discoveries at Prince Regent River described—Explained—Parvati—Siva—Mr. W. Froggatt's discoveries—NAURI—Hand-prints—Figures—Cave-paintings in New South Wales and at Billiminah Creek, Victoria—Sample of work at Billiminah Creek—Rock-carvings near Sydney—The Australian muse—Corroborees

CHAPTER XI

SORCERY, SUPERSTITIONS, RELIGION

Sorcery, superstitions, religion—The bane of sorcery—Native magicians or doctors—Their professed powers—Native phlebotomy—The rainbow—Spells—Names of deceased persons—Sacred pebbles—Ghosts—Ancient heroes—Deities

CHAPTER XII

AUSTRALIAN LANGUAGES

Introduction to Australian languages—Bleek's classification—The writer's classification—Fundamental principle of word-structure agglutination—Phonic system—Etymology—Formation of compound words—Kaiap, miowera, koonawara, koondooloo, kangaroo, kagurrin (name of laughing-jackass), wagan (name of crow), bomerang—Words—Particles—Noun—Number—Gender – Adjective — Numerals — Table showing relation of numerals—Pronoun and table—Prepositions and conjunctions—Verb

Pp. 149—174

CHAPTER XIII

OUTLINES OF GRAMMAR

Grammatical sketch of Tasmanian and of five Australian dialects representing the linguistic classes—Tasmanian—Wimmera, Victoria—Kabi, Queensland—Specimen in Kabi, with translation—West Australia—Diyeri, South Australia—Macdonnell Ranges, Central Australia .   .   Pp. 175—204

ILLUSTRATIONS

EAGLEHAWK AND CROW

A STUDY OF THE AUSTRALIAN ABORIGINES

CHAPTER I

THE ORIGIN OF THE AUSTRALIAN RACE

Origin, old designation Papuans—Dr. Lesson's theory—M. de Quatrefages'—Dr. Latham's view—Mr. E. M. Curr's—Dr. Topinard's—The writer's, first inhabitants Papuan—Dravidian immigration—Malay immigration.

IN entering upon a study of the Australian aborigines, the question "Who are they?" meets one upon the very threshold. Is the common belief correct that the lowly barbarians whose last vestiges are now rapidly melting away are the real indigenes? or is their being so called the result of hasty careless observation and imperfect knowledge? To call them the aborigines is convenient, especially as they have from their first appearance in history been so called, and as at a first view they seem homogeneous and without rival claimants to the distinction; but it will be easy to prove that the title does not strictly belong to them.

At the time when the Australian continent was known as New Holland, its inhabitants were loosely designated Papuans. In an old ethnological atlas in my possession they are classed among the Malays. After British settlement, observers among the colonists, by comparing the natives with typical Melanesians, could readily perceive very marked physiological differences, and

A

some colonial writers hit upon the hypothesis that the Australians were of mixed Papuan and Malay blood. The evidence in support was only the geographical position of Australia and the physical features of its people superficially scanned, and was so slight as to leave the allegation little more than a bare assumption. How, when, or where the fusion took place, if not insoluble, was not attempted to be solved.

The quite recent theory of Dr. Lesson, clearly and confidently stated, is almost identical with this dimly conceived one, but is better substantiated.· On physiological grounds Dr. Lesson* denies that the Australians have anything in common with the people of India, and he argues that in Australia and Tasmania three different races have combined, two of these being black, the other light brown or yellow (*jaune*). One of the black races was of short stature and brachycephalic or mesaticephalic, the other tall and dolichocephalic, while the third or yellow race was hypodolichocephalic. The Tasmanians he regards as the issue of the two first, the Australians of the two last. The brachycephalic race he identifies with the Negrito, the dolichocephalic with the Papuan, and the fair race with the Malay. His conclusion is based almost exclusively upon premises derived from craniometry, which, according to Huxley, is of little or no value for determining racial origin. The craniometrical difference, however, is very marked. There are three skulls of Tasmanian aborigines in the museum at Launceston. Two belonged to men of short stature; the third belonged to an aboriginal criminal whose height was over six feet. Compared by simple inspection, the last differs strikingly from the other two, being flattened at the sides and singularly elongated from front to back.

M. de Quatrefages held that the Tasmanians were a pure distinct race. A careful study of what is preserved of the Tasmanian language suggests that, although phonologically it is uniform, there are some indications that a close analysis might resolve its constituents into two etymological elements; and this may yet be done without proving that the original possessors of these elements belonged to different races. By combining the craniometrical and philological evidence, a good

---

* Dr. A. Lesson, "Les Polynésiens," Paris, 1880, vol. i. p. 104.

deal of support is given to Dr. Lesson's view that the Tasmanians were sprung from two dark races; but as this conclusion is uncertain, little or no stress will be laid upon it in this work.

The propinquity of Tasmania to the mainland naturally suggests the inference that both regions were at first peopled by the same race. Many accept this view off-hand without being aware how serious the objections are which it raises. Mr. Davies, quoted by Mr. R. Brough Smyth,* indicates King George Sound as the part of Australia whence the Tasmanians set forth, but no proof is given. Dr. Latham mentions the same theory, and gives some glossarial affinities, the validity of which as evidence the late Mr. E. M. Curr,† in his work "The Australian Race," very severely shakes, showing that a number of the words compared are not authenticated, and that of authenticated words only one of those given by Latham is represented in both Australian and Tasmanian speech. But Dr. Latham himself pronounces against concluding close relationship of races from contiguity of their abodes, and, by suggesting stronger affinities between the New Caledonian and Tasmanian tongues than between Australian and Tasmanian, leads his readers to prefer thinking that the migration to Tasmania had come by way of New Caledonia rather than from the mainland.

That a true relationship subsists between the Australians and the Dravidians of India is now admitted by various capable investigators on grounds too firm to be successfully controverted, as I cannot help thinking, notwithstanding Dr. F. Müller's stout assertions to the contrary.

One of the latest theories of the origin of the Australians is that advanced by Mr. E. M. Curr. He follows Mr. Hyde Clarke in citing resemblances between African and Australian words. Mr. Curr‡ concludes that the Australians and Tasmanians were respectively distinct offshoots from the African race; that the present occupants of Australia are its aborigines, and are so homogeneous that the founders of the race may all have arrived in the one canoe. He is at great pains to prove that Australia was never inhabited by the same race as the Tasmanians sprang

---

* "Aborigines of Victoria," Introd. p. lxx.
† "The Australian Race," vol. iii. pp. 600 *et seq.*
‡ *Ibid.* vol. i. p. 189; vol. iii. p. 604.

from, that in fact the Australians and Tasmanians have never met since first one of the branches was severed from the parent Negro stock.

The conclusions of Dr. Topinard,* derived mainly from a study of physical and physiological characters, while, as regards the mongrel constitution of the Australian race, corroborative of the theory of origin here enunciated, confess to the presence of such great difficulties on the question of origin, and show so much perplexity and uncertainty, that they may well be quoted as a warrant for marshalling such a mass of evidence as follows in support of the original occupation of Australia by the ancestors of the Tasmanians. He says " the *Tasmanian type* is separated in a most remarkable manner from all the neighbouring types, negroes or others." The Tasmanians are " absolutely *sui generis*." Some skulls appeared to be the product of a cross between the Melanesian and the Polynesian, but the Tasmanians " had a special physiognomy of their own." " The *Australian type* is no less paradoxical." " Is the Australian type a pure one ? " he asks. " We thought that before the present race of Australians there must have existed on their continent a race much inferior still, of whom the individuals with woolly hair and the ugly, deformed tribes were the descendants." " It is clear that the Australians might very well be the result of a cross between one race with smooth hair from some other place and a really negro or autochthonous race. The opinions expressed by Mr. Huxley are in harmony with this hypothesis. He says the Australians are identical with the ancient inhabitants of the Deccan." On the other hand, Topinard thinks that the few examples of woolly hair in the north " might be accounted for by the immigration of Papuans from New Guinea, and in the south by the passage over to the other side of Behring (*sic* for Bass) Strait of some Tasmanians to the continent." " We are still in ignorance as to whether the present Australian race took its origin on the spot with the characters that we admit as belonging to it, or whether, on the contrary, it was altogether constituted in Asia, or whether it is a cross race ; and in that case, of what elements it is composed." He has, therefore, no thought of the Tasmanians as the autoch-

* Dr. Paul Topinard, "Anthropology" (English Trans.). London : Chapman and Hall, 1890, pp. 500-505.

thones of Australia.   He leaves the origin an open question.

The present writer entertains the hope that this work will contribute to its solution. Topinard considers that no fewer than seven races of India and one of Ceylon are identical with the Australian, a most valuable support to the hypothesis presented here.

The theory which the writer enunciates accounts for the difficulties which give rise to these divergent views, and may be stated briefly as follows: Australia was first occupied by a people, a branch of the Papuan* family, and closely related to the Negroes. They came from the north, in all likelihood from New Guinea, but whether from there or any other island of the Eastern Archipelago is a matter of indifference and impossible to decide, as probably at the time of their arrival the islands to the north were all inhabited by people of the same blood. These first-comers, the veritable Australian aborigines, occupied all the continent, and having spread right across to the southern shores, they crossed what is now Bass Strait, but which at that distant date may have been dry land, and their migration terminated in Tasmania.

Then followed one invasion, if not two, by hostile people. The un-Papuan element now discernible in the Australian race is not the trace of one pure race, but is composite, the constituents being Dravidian* and Malay* blood. Of these the Dravidian was the first to arrive, the Malay coming later, and in a

* The three racial terms employed by the writer as distinguishing the constituent elements of the Australian race may be explained here once for all. Papuan is applied not in its narrowest application (dark New Guinean), but as the equivalent of Melanesian, and is meant to include the Tasmanian Aborigines as the vanguard of the race in the south. The writer is not aware that an absolute necessity exists for separating the Tasmanian from the Papuan, as Topinard does, making it a collateral distinct branch of the Negro family. Hence the Tasmanian Papuans are invariably referred to in this volume as the substratum of the present Australian race. That in them there may be a strain of Negrito blood is not questioned ; on the contrary, I incline to that opinion. Dravidian is not to be understood as indicating the direct descent of Australians from Dravidians (or Dravirians), but rather that one strong strain of the Australian people is of common origin with the Dravidians of India and their congeners. Malay refers generally to the people of that race to the north of Australia without distinguishing nationality. Proof will be given of intimate connection between Sumatra and the north-west of Australia. The superior physique of the Battaks may account for that of the natives in the north of Australia.

desultory way by detachments at irregular intervals. It is more convenient than accurate to designate one of these components as Dravidian; it would be more precise to speak of it of the same stock as the Dravidian. There are features observable in Australian marriage laws and indelibly fixed in Australian language which attest a real affinity between the Australians and the people of Southern and Central India. The different batches of invaders may have had different landing-places. Mainly from linguistic evidence I incline to think that the people, who for convenience may be called Dravidians, first touched on the north-east coast of Queensland. It seems to me that this ingredient of the population came not in one boatload, but in an unintermittent stream for many years, probably being forced southwards by the attacks of a more powerful race. Coming as a later off-shoot from the first home of humanity, this invading band was of higher intelligence and better equipped for conflict than the indigenes of Australia. Physically they were more lithe and wiry and of taller stature. They were lighter in colour, though a dark race, less hirsute, and the hair of their head was perfectly straight. Their language was dissimilar in phonology, and differed greatly in vocabulary from that of the indigenes. There is a natural highway easily traversed across Australia from the north-east to the south and south-west, by first ascending the rivers on the north-eastern watershed, and then descending those on the southern watershed until they converge about Lake Eyre. If we suppose the Dravidian invaders to have gained the extreme north-east coast of Queensland, thence they would rapidly pour south-westward in a strong stream, fighting their way with the aboriginal population, part of which they would absorb—chiefly by the capture of women—part they would destroy, the remainder would keep retiring. The stream of invasion would here and there send forth branches which, reaching the coast at various points, would rebound and eddy backwards.

As regards Malay incursions, while there may have been a continuous intercourse between Malays and Australians on the north chiefly to the west of the Gulf, there are not wanting indications of occasional descents of Malay parties—even on the east coast —forming, if not colonies, at least centres of influence, which have left unquestionable traces on the Australian language.

The theory of occupation which I have sketched differs widely from that propounded by Mr. Curr, who supposes that one boatload of people might have been the progenitors of the whole race. Following Mr. E. J. Eyre, he assumes that the landing was made near Port Darwin, and that afterwards the aborigines were by pressure of circumstances divided into three main advancing lines, two taking the seaboard in opposite directions, and the third penetrating into the interior, all three meeting again on the south and south-east coast.

Mr. Bonwick's theory differs *toto cœlo* from Mr. Curr's. He posits the existence of a great southern continent as Dumont d'Urville did, only Mr. Bonwick's continent would require to be more extensive, girdling almost the whole Southern Hemisphere, embracing the West Indies on the one side and the Chatham Islands on the other. Over all this vast continent people of Negro blood ranged, and were finally separated and isolated by a general depression and submergence of most of the land in the ocean. Both the Papuan and Australian branches of the Negro family advanced from west to east, reaching their permanent home at a time before Tasmania was sundered from the mainland. *The Papuans may have moved freely between both lands*, but the Australians touched on the south-west of the mainland and spread northwards and eastwards. Mr. Curr and Mr. Bonwick are almost diametrically opposed as to direction of settlement.

We may be disposed willingly enough to adopt Mr. Hooker's theory, based on botanical considerations, and called in by Mr. Bonwick as corroborative testimony, that the southern continent was once of much greater area than it now is (Mr. Wallace holds a similar opinion *), but it can only be a last resort to account for the distribution of races by the submergence of hypothetical regions. It will be conclusively shown in this volume that the main stream of population entered Australia on the north-east and crossed in a south-westerly direction.

* " The Malay Archipelago," p. 593.

CHAPTER II

THE INDIGENES OF AUSTRALIA PAPUAN

The Primitive Papuan base—Evidence from physiology contrast between Australians and Tasmanians—Statements *re* mixture of races: Mr. Jardine's, Rev. George Taplin's—Negro appearance of natives in the west—Appearance of blacks in north Australia—Shape of nose—Argument from mythology and tradition—Eagle and crow—Eagle and mopoke—Mr. McLennan on traditions of primitive man—Scripture metaphors—Heraldry—Australian classes, eaglehawk and crow—Mr. A. W. Howitt on classes—Eaglehawk and little owl—Apotheosis of heroes—Argument from implements—Stone tools—Clubs—Climbing-ropes—Argument from customs—Argument from language—Common phonology—Papuan lingual traces in Australia—Tasmanian speech crossed over from Victoria—Numerals—Idioms—Pronouns—Analogies—Tasmanian and New Caledonian.

HAVING expressed the conviction that the aborigines of Australia were Papuan, and that they were the ancestors of the Tasmanian race so recently extinct, I now propose to verify this hypothesis by presenting converging lines of cumulative evidence. There are proofs adducible from physiology, mythology, implements, customs and language, some more decisive and striking than others, but when combined so varied and powerful as, I think, to render my position incontestable.

EVIDENCE FROM PHYSIOLOGY.

The argument most obvious and first suggested for identity of origin of two nations is contiguity of habitation, and if considerations were just as favourable to the conclusions that the Tasmanians had sprung from the mainland as from any other place, proximity might be called in to turn the balance. Proximity of abode is here mentioned as a favourable presupposition to racial affinity, and with this brief notice it can be passed over as unnecessary argument and open to the objection of being sometimes misleading.

With Peron and many others, the most powerful argument against deriving the Tasmanians from their neighbours across the Strait has been the extraordinary difference in appearance. Descriptions of the physique of the Tasmanians vary exceedingly, so much indeed that it might easily be imagined that the writers had seen people of races physically as unlike as the Kafirs and the Bushmen. But similar divergence of impression is to be found regarding the appearance of the Australians. One writer will describe them as emaciated, undersized, disproportioned creatures hardly human; another will describe them as light but muscular, firmly knit men about as tall as Europeans. Much depends upon the subjective standard of comparison, the tribe met with, the number and circumstances of the individuals seen, and the season of the year.

By a most careful comparison of various accounts of both races, and judging from personal observation of the Australians, I am convinced that the obvious physical differences narrow down to these:[*] that as compared with the natives of the continent the islanders were on the average of shorter stature, of slightly darker complexion, and had hair of very different quality. The objections which these differences raise to community of origin dissolve in the prospect to those who accept the view, that upon the aboriginal Australian stock there was grafted a strong Malayo-Dravidian shoot, for the effect of this graft is of itself sufficient to explain the ultimate divergence of feature.

The average height of the Australian male may be set down as 5 ft. 5 in. or 5 ft. 6 in., while that of the Tasmanian was only about 5 ft. 2 in. to 5 ft. 5 in.[†] Any one who has seen much of the Australian blacks cannot have failed to observe the great disparity in stature to be found among them, whether as comparing together individuals of one tribe or individuals of different communities, their height as well as general physique being dependent upon descent, climate, food-supply. I remember meeting with a tribe on the Nogoa River in Queensland, which seemed to me a remarkably fine body of people, both for stature and for strength of build surpassing any natives I had seen else-

---

[*] The object here is mainly to refute Mr. Curr's arguments for holding that the Australians and Tasmanians were two absolutely distinct races, hence the peculiar cranial differences are not referred to.

[†] Bonwick, "Daily Life of the Tasmanians," p. 119.

where. We are told also that in some parts in the extreme north the natives are conspicuous for their height.*

When communities are examined in detail, it is found that a man's height will range from about 5 ft. to about 6 ft. 1 in. in the same tribe. I recall a singular example of this extreme difference in the case of two brothers, sons of the same mother at least. They belonged to the Kabi tribe, occupying the head waters of the Mary River, in Queensland. The younger brother, named Kilkaibriu, became a strapping fellow of about 6 ft., the elder brother, Kagariu, was very little over 5 ft. high, and about as unprepossessing, from a European point of view, as it were possible to conceive, in which respect no Tasmanian could surpass him. He came into public notice as Johnny Campbell, the bush-ranger, and after a singularly daring and villainous career, ended his days on the gibbet at Brisbane. Taken by itself, this great diversity of stature among Australians might only be regarded as something abnormal, but when joined with other evidence does it not form a link in the chain upon which hangs the hypothesis of their descent here advanced? The offspring of a union between the Australian Papuans and people of greater stature would probably exceed in height the pure aborigines represented by the Tasmanians, and among posterity sprung from two or three distinct races differing from each other in average stature, uniformity of height would hardly be expected, so that fact and theory coincide in this particular.

It is freely admitted that the inhabitants of the island differed physically from those of the mainland in important particulars; the difference, however, may be easily exaggerated. Strange that the colour of the Tasmanians should already be a matter of dispute. Mr. E. M. Curr † describes it as a sooty-black. A like opinion is given by Mr. Jas. Barnard, who had seen them, and calls their colour bluish-black.‡ The busts in the Melbourne Public Library are almost jet-black. Topinard § describes it as a chocolate-black, which corresponds fairly with

* The Woolna tribe are described as a fine race. My friend, Mr. Joseph Bradshaw, saw two women who were each 5 ft. 10 in. in height, and a young man 6 ft. 4½ in.

† "The Australian Race," vol. iii. p. 603.

‡ "Report of the Australian Association for the Advancement of Science," Melbourne, 1890, p. 598.

§ "Anthropology," p. 501.

the complexion of the natives in the painting by Mr. Dowling in the Launceston Museum. Mr. Dowling was a high-class painter, and as he painted from life I think his representation may be relied upon. Following him therefore, the contrast alleged by Mr. Curr between the sooty-black skin of the Tasmanians and the brownish-black of the Australians entirely disappears. The skin of both was brownish-black but uniform in the Tasmanians, and with marked differences of shade in the Australians, sometimes being copper-coloured, especially in children. Along the maritime lands of Australia on the east, south, and west, the colour in many cases is very dark.

The people of both nations had luxuriant heads of hair. Some have called the hair of the Tasmanians woolly, others deny that it could properly be so called, and aver that it was rather excessively curly. The hair of the head was very abundant and generally grew in long thin ringlets. The hair of the continental people is, on the contrary, mostly wavy on the head, but often straight, and occasionally so curly as to resemble woolly hair, while the beard has invariably a great tendency to curl. As telling against the common origin of these peoples, a strong point is made of the difference in the quality of the hair. Mr. Curr has said * of the Australians that their hair is "sometimes straight and at others wavy, but never woolly," and in the next sentence that the hair of the Tasmanians was woolly.

As a matter of fact, neither race had the hair woolly in the same sense as the Negro's hair is woolly, and yet the Tasmanians might be called in a sense woolly (or wool-like), and there are cases where a kind of woolly hair has been noticed among the Australians. To corroborate the latter statement I have only to refer to Mr. Curr's own work.† Of a tribe of blacks in the Bunya Mountains, the present writer, who contributed the account, says that there were one or two cases of woolly hair. The hair of one of them, named Warun, was so woolly that he used to be teased in consequence and nicknamed "Monkey" (sheep) and "Wool," much to his vexation.

Mr. Jardine ‡ (the explorer, I presume) speaks of two types of Australians, one approaching a copper-colour, the other black.

---

* "The Australian Race," vol. iii. p. 603. † *Ibid.* p. 153.
‡ Mr. R. Brough Smyth's "The Aborigines of Victoria," vol. i. p. 19.

He says the true Australian aborigines are perfectly black, with generally *woolly heads of hair*. He speaks also of features of a strong Jewish cast, about which more will be said below. Major Mitchell[*] saw some natives with a sort of woolly hair, and Mr. Stanbridge speaks[†] of isolated cases of woolly hair among the men. By the courtesy of a friend I have in my possession the photograph of a black boy whose hair was of the quality generally called woolly; his name was Wellington, and he belonged to the Culgoa River, New South Wales.

In his "Daily Life of the Tasmanians," Mr. Bonwick says,[‡] regarding them, that their hair was not woolly nor like that of the Negro. He cites the opinion of Dr. Pruner Bey, that two specimens examined resembled the hair of the New Irelanders. If Mr. Curr can hold that, notwithstanding the straightness of his hair, "the Australian is by descent a Negro with a strong cross in him of some other race," there should be no difficulty, on the score of difference of hair, in the way of our regarding him as descended from the Austral Papuan or indigenous Australian, with a strong cross of two other races, both straight-haired.

The opinion of the Rev. Geo. Taplin, an observer of large experience, is very noteworthy. He says,[§] "there is a remarkable difference in colour and cast of features; some natives have light complexions, straight hair, and a Malay countenance, while others have curly hair, are very black, and have the features Papuan. *It is therefore probable that there are two races of aborigines.*" My own theory was formed before I had read this; and besides, Mr. Taplin merely reiterates a supposition based certainly upon personal experience, but already propounded by earlier New South Wales writers,[||] and, apart from difference of appearance just quoted, he puts forth no proof of his statement.

The conclusion of so well qualified an authority that there are probably two races of aborigines in South Australia is in direct conflict with that of Mr. Huxley, who thinks that the natives of the southern and western portions of Australia are

* Mr. R. Brough Smyth's "The Aborigines of Victoria," vol. i. p. 18.
† *Ibid.* p. 15.　　　　　‡ "Daily Life of the Tasmanians," p. 106.
§ "Native Tribes of South Australia," p. 129.
|| Mr. Taplin regarded the Narrinyeri as descended from Polynesians and Papuans, and may have been the first to propound this special derivation.

WELLINGTON

MILROY STATION, CULGOA RIVER, N.S.W.

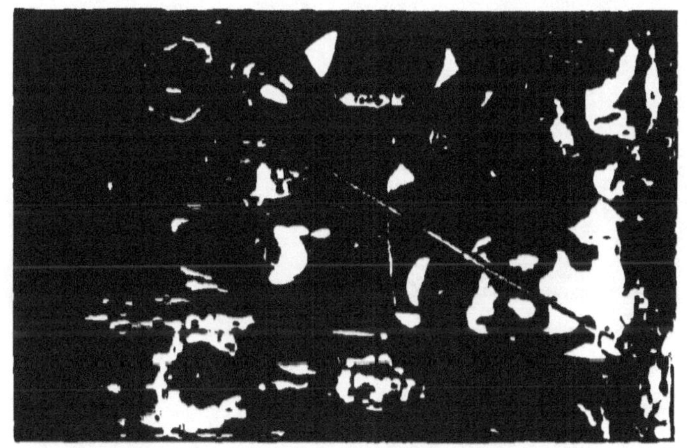

the most homogeneous of all savages.   Observation is certainly against Mr. Huxley, with whose opinion the statement of Dr. A. Lesson * may be compared, "the individual variations are too great, the study of the crania shows typical differences too accentuated for it to be possible to admit the unity and purity of the Australians."

The occurrence of strongly contrasted complexions, copper and almost jet black in the same tribe, is exceedingly common. Some of the fairer skins are accompanied by light-coloured hair, whether faded or natural.   At Beemery Station, between Bourke and Brewarrina, the family of the leading black were very fair and had long straw-coloured hair.   I have heard of similar cases elsewhere, and have known one or two in southern Queensland.

Mr. Bouwick † quotes Mr. Earl as saying, regarding Coburg Peninsula, in the north-west of Australia, "the aboriginal inhabitants of this part of Australia very closely resemble the Papuans of New Guinea, or, which is almost the same thing, the aborigines of Van Diemen's Land"; and on the next page Mr. Oldfield is credited with stating that the Papuan race is still showing through the Australian in a part of West Australia; "the tribes," he says, "inhabiting the country from Murchison River to Shark's Bay possess more characteristics of the Negro family than the aborigines of any part of Australia."   To the above evidence, attesting the greater prominence of Papuan characters in West Australia, let the following be added to show the existence of a decided Papuan fringe, at least on the south-eastern and western coasts, with a departure from it landwards and in the north.

Of some fishing tribes, Mr. Curr says ‡ that they have very frizzy hair.   Mr. A. W. Howitt, speaking of Cooper's Creek blacks, says § "the aborigines do not differ much in appearance from the coast blacks, but their hair is straighter, and I think they are slighter in build; the curly hair so often seen on the Darling and Murray and elsewhere is not common."

Mr. Curr quotes Mr. Paul Foelsche as saying, respecting the blacks of Northern Australia: "The majority of the men are

* Dr. Lesson, "Les Polynésiens," Paris, 1880, vol. i. p. 104.
† "Daily Life of the Tasmanians," p. 262.
‡ "The Australian Race," vol. i. p. 39.
§ Mr. R. Brough Smyth's "The Aborigines of Victoria," vol. ii. p. 301.

well built, but the skin is smooth and the stray covering of hair all over the body, so often met with in the south, is almost absent on the north coast . . . . the growth of hair on the face is very scanty."[*] As the character of the present Australian's hair is usually wavy, often curly and sometimes woolly, and as there is a special tendency to waviness towards the seaboard on the east, south and west, these considerations encourage rather than forbid the belief that the origin of the Australians was as here maintained.

I shall now conclude the argument from physiology by adducing the evidence which may be called *nasal*, and which, as might naturally be expected, is by no means microscopical. A feature common to the Papuans, Australians, and Tasmanians is a hooked nose. Mr. Wallace, who has studied carefully the Papuan and the Malay races, says[†] "the most universal character of the Papuan race is to have the nose prominent and large with the apex produced downwards." According to Mr. Jardine's description of the true aboriginal features cited above, the nose is that of the Papuans. The Rev. W. Ridley speaks of having met with the Jewish nose among the Australian blacks. The fact is so obvious to any one who has seen many of the natives, as not to require to be pointed out, much less supported by quotations. And Mr. Backhouse observed[‡] among the Tasmanian captives in Flinders Island, one especially whose features had a Jewish cast, and reminded him of the popular picture of Abraham. So that, besides the resemblances already noted, the Australians and Tasmanians are related by the family likeness of the Jewish Papuan nose.

THE ARGUMENT FROM MYTHOLOGY AND TRADITION.

Some myths have been collected, chiefly in Victoria, which at first appear to be wild nonsensical fancies, but which are capable of a beautiful and rational interpretation, and receive a special value when light is thrown upon them by the theory of pre-occupation of the country by a distinct race. While on this point it will be necessary for me to quote freely from Mr. Smyth's "Aborigines of Victoria," vol. iii., beginning at p. 423.

---

[*] Mr. Curr's "The Australian Race," vol. i. p. 248.
[†] "The Malay Archipelago," p. 590.
[‡] West's "History of Tasmania," vol. ii. p. 78.

The aborigines of the northern parts of Victoria believe that the beings who created all things had severally the form of the crow and the eagle. There had been constant warfare between these two beings, but peace was made at length. They agreed that the Murray blacks should be divided into two classes, the Mokwarra (spelt variously) or Eaglehawk, and the Kilparra or Crow. The conflict had been maintained with great vigour for a length of time, the crow taking every advantage of his nobler foe, but the latter generally had ample revenge. Out of their enmities and final agreement arose the two classes as has been said, and thence a law regulating marriages between these classes.

The Melbourne blacks say that Pundjel made of clay two males. He took stringy bark from the tree, made hair of it, and placed it on their heads, on one straight hair and on the other curled hair. The man with the straight hair he called Ber-rook Boorn, the man with the curled hair Koo-kin Ber-rook.

There is also a myth about Bundjel (or Pundjel), the first man, and Karween (the second man), whom Bundjel made. They quarrelled about wives, but Karween spoke to Waung the Crow and asked him to make a corroboree. And many crows came and they made a great light in the air and they sang. And then there was a fight with spears between Bundjel and Karween, the former being victor.

The following legend was current on the Murray. Before the earth was inhabited by the present existing race of black men, birds had possession of it. These birds had as much intelligence and wisdom as the blacks, nay, some say that they were altogether wiser and more skilful. The eaglehawk seems to have been the chief among the birds, and next to him in authority was the crow.

The progenitors of the existing tribes, whether birds, or beasts, or men, were set in the sky and made to shine as stars if the deeds they had done were mighty. The eagle is now the planet Mars, and justly so, because he was much given to fighting; the crow is also a star.

The Murray blacks have it that the crow killed the son of the eagle. This made the eagle very angry, so he set a trap for the crow, caught him and killed him, but the crow came to life again and disappeared.

The Gippsland blacks vary the legend by saying that the eagle left his son in charge of the mopoke while he himself went hunting. The mopoke sewed the eaglet up in a bag and left him. The eagle was irate, got the mopoke enclosed in the cavity of a hollow tree, whence he was able to escape only by breaking his leg and using the bone of it to cut his way out. The eagle and the mopoke afterwards made a solemn agreement and treaty of peace, the conditions of which were as follows : the eagle should have the privilege of going up into the topmost boughs of the trees, so that he might from so great a height see better where kangaroos were feeding ; and the mopoke was to have the right of occupying holes of trees : thus ended the disputes between the eagle and the mopoke.

The Rev. Geo. Taplin relates some myths of the Narrinyeri in South Australia, similar to the above.* Nurundere was the wonderful god or chief of this tribe. When he and his followers came down the Murray they found the country around the lakes in possession of clans of blacks under Wyungare and Nepelle. The last two of these heroes were translated to heaven and became stars. There is also a legend of a fight about fish between the pelicans and the magpies, when the latter were rolled in the ashes of a fire they had made and became black. This myth, like those about birds narrated above, will bear a similar interpretation.

Now what is to be made out of these myths? Are they tales "told by an idiot and signifying nothing?" or are they confused evanescent echoes of a real past history? I take them to be the latter. Primitive man was fond of representing warfare carried on between beasts or birds endowed with human faculties, or between men and some of the lower animals, and men were united with beasts in all sorts of relations. A number of these relations are mentioned by Mr. McLennan;† such as the Minotaur and his parentage, Phorbas attaining the supremacy in Rhodes by freeing it of snakes, the conversion in Ægina of the ants into men, the Myrmidons; "and a score of suchlike facts." He asks what these relations meant, and suggests that among the Greeks there were tribes with *totems*—Bull, Boar,

---

* "Native Tribes of South Australia," pp. 55-62.
† "Studies in Ancient History." London : Macmillan and Co., 1896 p. 227, *note*.

Lion, Snake, Ant, and Dragon tribes, just as there are tribes named after animals among the American Indians.

The prevalence of the designation of men by names of the lower animals is amply illustrated in the Old Testament scriptures. Take, for instance, the case of Jacob blessing his children,* where Judah is "a lion's whelp," Issachar "a strong ass," Dan "a serpent by the way, an adder in the path," Naphtali "a hind let loose," Benjamin "a ravening wolf." In the book of Daniel† the empires are typified by four beasts. There is also the common appellation for Egypt, "the dragon."‡

This ancient practice has been handed down to modern times in the heraldic bearings both of families and nations in civilised countries. The eagle has always been a choice crest, and it is scarcely matter of surprise that the king of birds, so swift and fearless, should be chosen as the emblem of a conquering people even in Australia.

Standing in close relation to these myths is the division of Australian communities into two classes, represented by the eaglehawk and the crow respectively, this dual division and particular representation occurring in Victoria, and extending with modifications into New South Wales and South Australia. In central and northern Victoria the eaglehawk and the crow are the only names of the two classes. Throughout much of the watershed of the Darling and the Murray, on the authority of Mr. A. W. Howitt,§ the eaglehawk is one of the primary class-names, the second name being usually the crow. In the Turra tribe in South Australia, bordering on the south-west of Victoria, the seal takes the place of the crow. Merung, *eaglehawk*, and Yukembruk, *crow*, are the two class-names on the Upper Murray and at Maneroo, New South Wales. Bearing upon this is the tradition of the blacks on the Lower Darling, first placed on record by Mr. C. G. N. Lockhart in his annual report to the Government of New South Wales in 1852 or 1853, and cited by Mr. Curr.‖ The tradition is that the first black man on the Darling had two wives, Kilparra and Mokwarra. The sons of the one married the daughters of the other, and the class-names were inherited from the mothers. At King George

* Genesis, xlix.   † Daniel, vii. 3.   ‡ Isaiah, li. 9.
§ "Kamilroi and Kurnai," p. 288.
‖ "The Australian Race," vol. ii. p. 165.

Sound, among a community of the Meenung blacks, the white cockatoo is substituted for the eaglehawk as one of the primary divisions, the crow being the other.*   In central Victoria, Bunjil (*eaglehawk*) was the name of a deity as well as a class-name. In Gippsland, Victoria, it was a title of respect applied to men. At the most easterly point of Australia, between the Albert and Tweed Rivers, the equivalent of "blacks" is Meebin, which also signifies eaglehawk; and even farther north still, the name Dippil is applied by the Rev. W. Ridley, who received it from Mr. Davies ("Darumboi"), as an equivalent for Kabi, the name of the blacks in that quarter; and Dippil is evidently the same word as dibbil (eaglehawk) of the Brisbane River blacks.   It is only natural to infer that these correspondences between the name of the race and that of the eaglehawk result from the fact that the dominant and predominant race was called after that bird.   "Among the Kurnai," writes Mr. A. W. Howitt,† "the eaglehawk is greatly reverenced; he is regarded as the type of the bold and sagacious hunter . . . . He figures in their tales in company with Ebing, the little owl.   Were it not too fanciful, we might see in the quarrels of Gwanumerong and Ebing a trace of the severance of the original community into two classes, or of a special disruption which may have impelled the Kurnai ancestry into Gippsland."   When the natives of one tribe or community all belong to one class, and those of another tribe belong to a different class (as in central and northern Victoria and elsewhere), surely we are justified in interpreting the mythical bird-warfare as referring to the classes, and therefore, necessarily, to the communities which bear the bird-names. The theory advanced here goes a step farther, identifying the two primary classes with two races; and if it be accepted, the strife is regarded as not merely inter-tribal but inter-racial. A hatred or dread of crows is evinced in places widely separate. In a note from Mr. Shearer, speaking of the tribe living between the Culgoa and Warrego Rivers, he says: "If they cut their hair, they are very particular about leaving it, for fear of the crows picking it up.   They suppose the hair on their head would turn to grass or sticks if the crows took it.   They have a great dread of crows.   If they see a flock making a noise they

* "The Australian Race," vol. i. p. 386.
† "Kamilroi and Kurnai," pp. 322-23.

are sure some other tribe are going to fight them or afflict them with some sickness."

When we take a conjunct view of the myths of the eaglehawk and the crow, the widespread currency and imperishable persistence of one or other of these names as applied to tribes or divisions of tribes, the Darling tradition of the aboriginal with the two wives, the persisting hatred and dread of crows, is there any better explanation of the facts possible than that the eaglehawk and the crow represent two distinct races of men which once contested for the possession of Australia, the taller, more powerful and more fierce " eaglehawk " race overcoming and in places exterminating the weaker, more scantily equipped sable " crows " ?   The struggle for supremacy began in the north and its last smouldering embers died out in Victoria, where traces of the once fierce fire have been left as clearly recognisable as the Victorian evidences of a former volcanic period, and a not inappropriate name, for the south-east of Australia at least, would be THE LAND OF THE EAGLEHAWK AND THE CROW.

The myths of Looern and Wiwonderrer suggest that they relate to untamable Papuans holding out for some time in the wildest parts of Victoria. Looern had his house at Wilson's Promontory. His country was that tract of heavily timbered ranges lying between the Promontory and Hoddle's Creek. Any who dared to penetrate this country without the permission of Looern died a death awful to contemplate, because the torments preceding death were indescribable. The myth of Wiwonderrer is briefly stated thus :* There is a range north-east of Western Port inhabited, the natives say, by an animal resembling a human being, but with a body as hard as a stone. He used to kill many blacks. He was supported by people of his own. The blacks would not visit this range on any account.

Mr. Stanbridge states that the Boorong tribe, who inhabit the Mallee country in the neighbourhood of Lake Tyrril, have preserved an account of the Nurrum-bung-utrias or old spirits, a people who formerly possessed their country and who had the knowledge of fire. This tribe imagined the star Canopus to be

---

* Mr. R. Brough Smyth's " The Aborigines of Victoria," vol. i. p. 453.

the male crow, the first to bring fire from space and to give it to themselves, before which they were without it.*

There is a great resemblance between the Victorian and Tasmanian legends of the origin of fire and the apotheosis of heroes. Thus, according to the Yarra blacks, Karakarook, a female, was the only one who could produce fire, and she is now the seven stars (the Pleiades presumably). There is another Victorian myth to the effect that Toordt and Trrar came from the sky to show the blacks where the crow (that hostile wicked crow) had hidden fire and returned to the sky again. Pundjel is said to have changed Toordt into Mars for his good deeds.

With the foregoing may be compared the legend of the Tasmanian Oyster Bay Tribe preserved by Dr. Milligan. Two strangers are said to have appeared suddenly and to have cast fire among the Tasmanians, and, as the legend goes, " these two are now in the clouds; in the clear night you see them like two stars (Castor and Pollux)." The resemblance between these Victorian and Tasmanian myths, little in itself, forms yet another link in the evidence for the relationship of the two races.

Mr. West observes† that a New Holland woman taken to Flinders remembered a tradition that her ancestors had driven out the original inhabitants, the fathers, it is conjectured, of the Tasmanians; but as the navigator could hardly be able to interchange ideas on such a subject with a native at that time, even with all the resources at his command, the story is of very little weight.

Against my interpretation of these bird-myths it may be urged with great show of reason that the most they can suggest in the way of ancient warfare is a feud between two clans having bird-totems, and that in the Australian communities there are always in the one tribe two classes at least, a circumstance favouring the presumption that a duality of classes existed among the race from which the aborigines are sprung for ages before Australia became their home.

But my theory is strongly corroborated by the system of classes which prevailed generally in Victoria and in the adjoining part of South Australia south of the Murray. Here (in Gipps-

* Mr. R. Brough Smyth's " The Aborigines of Victoria," vol. ii. p. 460.
† " History of Tasmania," vol. ii. p. 77.

land) there was the peculiarity of sex-totems, the origin of which would be explicable upon the supposition of wives retaining the name of the totem of their kin, their tribe or race being different from that of their husbands.   Here there was also the somewhat rare system of local or tribal totems with corresponding classes. All the native-born in the tribe or locality took the name of the father's class, say, *Eaglehawk;* the tribe was exogamous, wives being taken from, say, a *Crow* community, in which the same principle was acted on, the class-names being transposed.

The Narrinyeri, south of the mouth of the Murray, had eighteen such communities, each having its own totem and forming an exogamous class, the children taking the father's class-name, and thus perpetuating the territorial totem and class. Such a system is quite consistent with a racial dual division characterised by the names *Eaglehawk* and *Crow*—in fact, points to it.

And the more complicated systems of Queensland, with four clans and two phratries, can be explained as arising from the simpler Victorian usage.   The Queensland east coast systems may be represented briefly thus:

$$\text{Phratry I.}\begin{cases}\text{Class A.}\\\text{Class B.}\end{cases}\qquad\text{Phratry II.}\begin{cases}\text{Class X.}\\\text{Class Y.}\end{cases}$$

The identity of the phratries is still marked in some places by distinct names.   A and B do not intermarry, nor do X and Y. B marries X, the children are Y, and so on through all the possible combinations.   This gives a succession through females of XY XY *ad infinitum*, and on the other side of AB AB correspondingly.   Suppose that each phratry represents a fusion of two communities.   In one phratry there were the clans A and B, and if Victoria, as being the most primitive in language and most closely related to Tasmania, indicates the early type of community generally, the A and B classes or clans were each tribal or territorial.   A had one territory, B another; they cross-married; the descent, regarded through males, would run A' A'' A''' and B' B'' B''', but through females, AB AB AB.   The same order would prevail in phratry II. with X and Y classes. Then, if two compound communities, having lived apart from one another for many years, were to meet and become gradually fused, and if the clan-names of the women were to determine the style of nomenclature of the offspring, there would result exactly the system found along the Queensland coast from Brisbane to Mackay.

Amongst the Kabi, in the south of Queensland, a member of a clan of one phratry could marry into either clan of the other phratry.

Hence the Queensland system is easily explicable as a *natural development* of the Victorian, and the Victorian is not inconsistent with the theory of the coalescence of two originally distinct races recognised respectively as *Eaglehawk* and *Crow*, which names may have been those of their totems.

The theory here propounded of the origin of the classes being simple and natural, and supported by the class-systems of the most primitive Australian inhabitants (or at least those who retain most distinct marks of the autochthonous race), is surely much more reasonable than a theory which requires the formation of classes to be due to far-seeing deliberation on the part of savages, such foresight resulting in a complicated scheme.

But whether I am right or no in believing the names *Eaglehawk* and *Crow* to have designated two races, they certainly designated clans over so extensive an area that I am quite justified in adopting them as part of the title of this book.

THE ARGUMENT FROM IMPLEMENTS.

As compared with the implements and weapons of the continent, the paucity of these in the hands of the Tasmanians, the rudeness of the form and the inferiority of the workmanship, suggest a difference of descent in the makers. But the lower skill of the islanders may be easily accounted for by the supposition that their progenitors had already reached Tasmania before the better-equipped race had reached Victoria, and that after the first settlement of the island, which may have been made when it was much more accessible than now, no further communication took place with the mainland. It is hardly fair to compare the weapons of the Tasmanian with those of the Australian, and from their dissimilarity to deduce absence of racial affinity in the owners, for the isolation of the Tasmanians reduced them to dependence for advancement on a very limited number of minds, and they may have made little or no progress after they crossed Bass Strait, whereas their kin on the mainland were overwhelmed by a race bringing with them superior art, which, once introduced, only faint traces of the work of the primitive inhabitants might be expected to linger on. It is futile to ask whether all the Australian implements are represented in Tasmania. If the implements of Tasmania be also found in Australia, although of improved manufacture, that should be sufficient to justify the theory propounded here in so far as the argument from such belongings has any force. The fact that certain weapons of the continental natives are absent from the island forms part of Mr. E. M. Curr's reasons for supposing that the Tasmanians were not of Australian descent, a method of reasoning which would lead inevitably to the conclusion that some of the Australian tribes were not of Australian descent.

For instance, neither the shield nor boomerang were known to the Tasmanians, nor were their weapons ornamented. But this ignorance is exactly paralleled by a people on the mainland. In Mr. Curr's own work * we read that among the Wonunda Meening tribe of Eyre's Sand Patch, " Shields and boomerangs are unknown, and their weapons are unadorned with either carving or colouring." This tribe also resembles the Tasmanians in being without the usual message-stick. It is true that for arms the Tasmanian had only a plain spear and club, but these are universal in Australia, where the variety and more artistic make may be ascribed to the influx of a more advanced people and to the greater scope for and stimulus to invention on a territory so much more extensive and populous.

The club of the Tasmanians was pointed at both ends, and the part to be grasped was roughly notched so as to afford a secure hold for the hand.† This description would apply equally well to the common club or *kuthar* used by the blacks in southern Queensland, which was entirely destitute of ornament. I mention this locality particularly, because I have accurate knowledge of the fact stated, and not because the plain weapon was only in use there.

Mr. Curr does not credit the Tasmanians with the ownership of a tomahawk or stone axe, as others have done. They certainly had a stone cutting implement, call it what you like, some

* " The Australian Race," vol. i. pp. 395-96.
† Mr. R. Brough Smyth's " The Aborigines of Victoria," vol. ii. p. 400.

specimens being beautifully finished, as Mr. Brough Smyth testifies from inspection. It seems almost incredible, that after the lapse of so short a time we should be unable to determine for certain whether the tomahawks of the Tasmanians had handles or not. There is some strong evidence that they had. Thus, *e.g.*, while Mr. Gunn says,[*] " The tomahawks were held in the hand, and under no circumstances, as far as I know or can learn, were they ever fixed in any handle," a Mr. Rollings, in a letter addressed to Dr. Agnew, and dated May 5, 1873, says that in his youth he was constantly in the habit of seeing the aborigines of Tasmania and of mixing with them occasionally, and he affirms that their tomahawks had handles which were fastened to them in the same way as a blacksmith fastens a rod to chisels, being always well secured with the sinews of some animal.

But, even if it be conceded that the Tasmanians used their axes without handles, the admission does not in the least invalidate the present argument as to their origin, for we find that the natives of the northern tributaries of the river Darling do not in all cases attach handles to their stone hatchets, but may use them in the same manner as the Tasmanians used their rough stone tools.[†]

It is of more consequence to note the difference in the mode of forming the large stone tools. In Tasmania they were usually chipped to an edge, in Australia they were almost universally ground and polished. But even here exceptions in Australia indicate a former more primitive manufacture. The chipped stone tools of the Tasmanian are Palæolithic; the usual ground ones of the Australian are Neolithic; but while as a rule only the one kind (Palæolithic) is found in Tasmania, both kinds are found side by side on the mainland. The opinion commonly entertained that the Tasmanians had no stone implements ground to an edge must be erroneous, Dr. E. B. Tylor having got possession of genuine specimens thus finished.[‡] " If, therefore," says Mr. Brough Smyth,[§] " all the stone implements and weapons of the Australians be examined, one set might be put

---

[*] Mr. R. Brough Smyth's " The Aborigines of Victoria," p. 403.

[†] *Ibid.* vol. i. p. 55.

[‡] " Journ. Anthrop. Inst." vol. xxiv. p. 339.

[§] " The Aborigines of Victoria," Introd. p. lv.

apart and classed as the equivalents of those of the Palæolithic period of Europe, and another set as the equivalents of those of the Neolithic; a man of one tribe will have in his belt a tomahawk ground and highly polished over the whole of its surface, and not far distant from his country a people will use for tomahawks stones made by striking off flakes."

I cannot refrain from quoting here the same writer's conclusions based upon difference of arms used by the two peoples. "The character of the weapons," he says,[*] "made by the natives of Tasmania, the absence of ornament, their using their clubs as missiles and throwing stones at their enemies when all their clubs were hurled . . . indicated a condition so much lower than that of the Australians, that *one is not unwilling, with Dr. Latham, to seek in other lands than those from which Australia was peopled for their origin.*" It is a pity that such a conclusion should have been expressed in a book which must always remain an authority upon the Australian aborigines, because it is altogether unwarrantable, inasmuch as the various marks of inferiority which characterise the Tasmanians are found here and there on the mainland. For instance, it has been shown above that in certain parts of Australia the tomahawks are used without handles, and in other parts the shield and boomerang are unknown and the weapons unadorned.

Mr. Smyth assumes that the Australians *do not throw their clubs*, but they do. The club was the proper weapon of the Kabi tribe of Queensland (as of others, no doubt) for hunting the kangaroo, and they usually hurled it in the chase. And moreover, we are told that the natives of Cooper's Creek were in the habit of throwing stones in warfare. So that the logical conclusion to deduce from the arms of the Tasmanians is that they were of the same kind as those of the lowest of the Australians, and it is anything but illogical to infer that the autochthonous Australians once used exactly the same weapons and instruments as those of the islanders, but by circumstances which affected only the continent, the arms and implements there were almost universally improved.

One instrument, and a very important one, extensively used by the two nations has hitherto been overlooked as evidence of

[*] Mr. R. Brough Smyth's "The Aborigines of Victoria," vol. ii. p. 401.

their kinship—I refer to the rope for climbing trees. It is hardly a mere coincidence that this rare and most valuable device should be found on both sides of Bass Strait. The material of which the rope was made differed in different localities in both countries, but the mode of use and the skill of climbing by its aid were pretty much the same. Mr. West says[*] that by this means the Tasmanians ascended almost as quickly as with a ladder and came down more quickly. I have seen an agile black woman on the Bunya Mountains in Queensland walk up a tall, smooth, perpendicular tree by the aid of the rope at quite a military quick-march pace. In Tasmania the rope was made of kangaroo sinews or grass twisted, and handles were attached. At Twofold Bay, in New South Wales, the material of which it is made is the fibre of some vegetable, and here the rope is also provided with *wooden handles.*[†] In some parts of Victoria it is made of stringy bark. In the southeast of Queensland a tough vine is used, and I have even seen a very light iron-bark sapling improvised for a climbing-rope.

The Tasmanians had also baskets like those of the natives of the continent, and the ovens so common in Victoria are said to be found occasionally in Tasmania.[‡]

THE ARGUMENT FROM CUSTOMS.

When we compare the customs we find a very marked resemblance—in fact, it may be truthfully said that such customs as are universal in Australia were all followed in Tasmania. The dwellings of the two peoples were identical. Of the Tasmanians it is said,[§] "Their huts were of bark, half-circular, gathered at the top and supported by stakes." For houses they also made break-winds of boughs formed in the shape of a crescent with a fire burning in the open space in front, and near Pieman's River, on the west coast of Tasmania, "one tribe was discovered living in a village of bark huts or break-winds of a better description than usual."[||] These notices

---

[*] "History of Tasmania," vol. ii. p. 86.
[†] Mr. R. Brough Smyth's "The Aborigines of Victoria," vol. i. p. 151.
[‡] Mr. Bonwick's "Daily Life of the Tasmanians," p. 19.
[§] Mr. West's "History of Tasmania," p. 82.
[||] Mr. R. Brough Smyth's "The Aborigines of Victoria," vol. ii. p. 389.

form also a perfect description of the dwellings on the mainland.

The following practices were common to both peoples: initiatory rites to manhood, enforced abstinence from certain kinds of food, remedial bleeding, the wearing as charms the bones of deceased relatives, refraining from mentioning the names of the dead, laceration of the body by women in mourning, ornamental cicatrising of the bodies of young men, exogamy, polygamy, burial in hollow trees, accumulation of skulls in cemeteries, carrying of sacred stones for the injury of foes and the benefit of friends, the obtaining possession of an enemy's hair to cause his death, knocking out one or more of the front teeth, ornamentation of the body with charcoal, red ochre, and pipeclay; climbing trees by means of notches, and also of a climbing-rope; submitting to the penalty of receiving strokes from a club or casts of spears as expiation of offences against the tribe, making the women beasts of burden and generally ill-treating them, hereditary feuds, sketching living objects in charcoal, the hunting of kangaroos by firing the grass and intercepting retreat. This list of remarkable practices, identical in both countries, is surely sufficiently imposing to establish of itself a very intimate connection, if remote in time.

It is a matter of dispute whether the Tasmanians knew how to produce fire, but Mr. Davies states that he was informed that they obtained it by rubbing round rapidly in their hands a piece of hard pointed stick, the pointed end being inserted into a notch in another piece of dry wood.* And an ancient ex-bushranger told Mr. Bonwick† that to produce fire the natives got two pieces of grass-tree stem, the smaller piece having a hole in it. " Soft downy inner bark of trees was mixed with powdered charcoal and placed in the hole, and friction with the other stick ignited this and produced a flame." Exactly the same method was used on the mainland.

Mr. Curr denies‡ that the Tasmanians practised the corroboree, but there is abundant evidence that they did. Mr. Davies says that their chief amusement consisted in the

* Mr. R. Brough Smyth's " The Aborigines of Victoria," vol. ii. p. 408.
† Mr. Bonwick's " Daily Life of the Tasmanians," p. 20.
‡ " The Australian Race," vol. iii. p. 598.

corroborees or dances. Mr. Bonwick writes : " The corroboree in the Tasmanian woods was very similar to that of the Australians, being chiefly by moonlight, though by no means confined to that season. A great corroboree took place at the full moon of November each year."[*] And Mr. Hill's more precise description of their singing and dances is well worth noting. " They sang," he says, " all joining in concert, and with the sweetest harmony. They began, say in D or E, but swelling sweetly from note to note, and so gradually that it was a mere continuation of harmony; their dances are a mere wriggling motion of the hips and loins, obscene in the extreme." This description would apply exactly to some of the Australian corroborees, and the abominable motions in dancing are also precisely like what is common in Australia, and, so far as I have heard, without parallel elsewhere.

Another example of the invalidity of reasoning from the absence of certain practices in Tasmania that were found on the mainland is the following from Mr. Curr's in many respects most excellent work : " The Tasmanians," he says,[†] " neither skinned nor disembowelled animals before cooking, but laid them whole on the fire." In the same work we are told[‡] that the Muliarra tribe in Western Australia place the animal to be roasted on the fire whole, and take out the entrails when it has been partly cooked. He continues : " Fire was not made by friction of wood nor cannibalism nor circumcision practised." First-rate testimony has already been adduced to the knowledge possessed by the Tasmanians of producing fire by friction. If we affirm that they were not cannibals, we must base our opinion upon our ignorance rather than our knowledge ; but even if they were not, we find in this respect a likeness between them and certain Australian communities, as, for instance, a very low tribe at Eucla, in South Australia, among whom cannibalism was unknown.[§] The same statement holds good in respect of Australian tribes widely distant from this one, such as the tribe at the junction of the Lachlan and Murrumbidgee,[||] and the Murraworry tribe between the Warrego and Culgoa Rivers.[¶] That circumcision was not

* Mr. Bonwick's "Daily Life of the Tasmanians," p. 38.
† "The Australian Race," vol. iii. p. 598.
‡ *Ibid.* vol. i. p. 376.   § *Ibid.* vol. i. p. 402.
|| My informant is Mr. Humphry Davy.
¶ My informant is Mr. William Shearer.

observed in Tasmania is of no consequence to prove different
derivation of peoples, for if it did, the argument would recoil on
Mr.Curr by proving too much.   It would split up the inhabitants
of Australia into two races distinct in origin, for the observance
of circumcision in Australia is limited to the people of a broad
central belt crossing from north to south.   Farther on in this
memoir the partial distribution of circumcision in Australia will
be accounted for adequately.   It is very clear, therefore, that
the inferences based upon alleged dissimilarity of customs are
of little or no weight, and especially when the numerous and
striking points of similarity are taken into account.   We may,
however, with perfect fairness conclude that such peculiar practices as are common to the two nations have been inherited
from the primitive Papuan Australians.

THE ARGUMENT FROM LANGUAGE.

The last, and perhaps the most important, class of evidence
attesting the community of origin of the Tasmanian and, at any
rate, one element of the mainland race, is that offered by their
language. Upon careful inspection the Australian and Tasmanian
languages will be found to exhibit unmistakable resemblances not
alone in phonology and structure but also in a considerable
number of vocables.   When one who has been accustomed to the
dialects of southern Queensland and New South Wales begins to
study those of Victoria, he cannot help being struck with some
entirely new features distinguishing the last named.   The
Kamilroi, Wiradhuri, and allied dialects are singularly fluent and
melodious and free from harsh sounds. The initial and final
letters are very limited, and certain combinations of consonants
are avoided.   For instance, in these northern tongues no local
word begins with 'l,' only one or two words with 'r,' and only
an odd one, if any, ends with 'k'; whereas all over Victoria, and
extending along the Murrumbidgee into New South Wales, a
good many words begin with 'r,' and the initial 'l' and final 'k'
are quite common.   This is a most conspicuous difference, which
as you travel southward is met first about the Reed Beds on the
Lachlan a little above its junction with the Murrumbidgee.   If
a geologist, in tracing a bed of limestone, finds it suddenly transformed into marble, he is sure that metamorphic fires have been

at work; and just as reasonably does the philologist conclude
the former interference of a powerful disturbing cause when he
finds at a particular line a sudden change in the genius of a
language. The proximate cause of the difference just noted
appears to be a more decided residual Papuan element in
Victorian speech than in the dialects farther north. Of the
latter, let the Kabi dialect of Queensland, spoken in the Bunya
Mountains, stand as a special example. It has no word beginning
either with ' l ' or ' r.' Its terminal letters are limited to ' l,' ' m,'
' n,' ' r,' ' ng,' and vowels. In general it may be said that such
combinations as ' bl,' ' br,' ' gr,' common enough in Victoria, are
of very rare occurrence in the north. An examination of the
scanty remains of the Tasmanian speech shows that it is characterised by initial ' l,' initial ' r,' and final ' k '; ' kl,' ' pl,' and ' bl,'
' kr,' and ' dr,' ' rt,' and ' rk,' are common Tasmanian unions and
not infrequent in Victoria, while they are of comparatively rare
occurrence in other parts of the continent. Where, save in
Victoria, would such forms be found as ' grangurk,' *a hill;*
' ngurnduk,' *teeth;* ' kroombook,' *breasts;* 'kraigkrook,' *mosquito?*
with which compare Tasmanian ' crougana,' *aloft;* ' krangboo-
rack,' *ripe;* ' neoongyack,' *rage;* ' crackaneeack,' *ill.*

In fact, it is obvious beyond any question that, while we
discover positive Papuan (Tasmanian) lingual traces in most
parts of Australia, with slight exceptions they are more distinct
on the coast than inland, more strongly marked on the west
coast than on the east (north of Victoria), still more numerous
on and near the north coast, and they are most abundant and
most conspicuous in Victoria, proving without a doubt that the
Victorian dialects inherit a powerful base of the primitive Papuan
or Tasmanian language, and leading to the conclusion that *the
Tasmanian speech crossed over from Victoria.*

The most remarkable negative features of both the Australian
and Tasmanian tongues are the absence of sibilants and of the
decided palatal ' ch ' and soft ' g.' The fact is not overlooked
that some writers have introduced an occasional ' s ' or ' z ' into
Australian orthography, as others have into Tasmanian, but the
rarity of these in both cases is so extreme as to be phenomenal,
and sometimes rather attributable to the ear of the hearer than
to the tongue of the speaker. I am aware that many spell
Australian words with ' ch ' and English ' j ' or soft ' g,' but the

sounds thus represented would, it seems to me, be more perfectly written as 'ty' or 'dy,' the 'y' having its consonantal English value.   Thus, instead of 'cha,' it would be more like the native pronunciation to write 'tya,' alternative modes often met with, and instead of 'polaich' or 'polaitch,' it would be more accurate to write 'polaity,' the 't' and 'y' coalescing.   Indeed, the Adelaide 'parlaitye,' *two*, corresponds to Victorian 'polaitch.' The only sound in the Tasmanian speech which with any show of reason can be said to be wanting in the Australian is the guttural 'ch,' to which Mr. Curr adds the French 'u.'   As these are the only two sounds adduced by Mr. Curr as indicating dissimilarity of phonology and forming part of the evidence of alienation of blood, it may be observed in reply that the French 'u' to most English ears varies with the ear, and that among the Australian aborigines there are peculiar modes of enunciating certain obscure sounds which have never been represented on paper. I have heard in Queensland a terminal combination of 'iu,' which some would call a French 'u.'   But so subtle a variation in the pronunciation of a vowel might only be provincial, just as in some parts of the Lowlands of Scotland the French 'u' is found and not in other parts, although the people throughout are of the very same stock and speak the same language.

The argument based on the absence of this sound in Australia is completely nullified by the statement of Mr. Schürmann[*] that "*the aboriginal language requires sounds like the French 'u' or German 'ü.'*"   An opinion evidently shared by another German, a member of the Roman Catholic Mission at Port Darwin, who has favoured me with a vocabulary of the Larrikeeya tribe, in which he employs the German vowels 'ö' and 'ü.'   The guttural 'ch' is certainly very rare in Australia, but singularly (perhaps I should say naturally) enough we are told that it is used in Victoria and on the south-east border of South Australia.   On the Upper Richardson 'h,' a closely related guttural, was sounded[†] clearly and sharply like 'r.'   Mr. Hartmann says of the Victorians that the 'h' of the third person plural scarcely expresses the sound it is meant to express, the 'ch' should be pronounced as the German 'ch' in *ich, mich, sich;* and the Rev. Geo. Taplin says that 'h' was

---

[*] "Parnkalla Vocabulary," p. 2.

[†] Mr. R. Brough Smyth's "The Aborigines of Victoria," vol. ii. p. 3.

sounded clearly and sharply among the Narrinyeri on the Murray River bordering on Victoria.

Unfortunately for a comparison of syntax and general structure, no Tasmanian grammar was ever compiled, so that we can only base inductions upon the few brief dialogues and meagre vocabularies that have been preserved. However, from these we glean the following points of resemblance to Australian syntax, ideology, and word structure. The Tasmanians modified by post-positions, the usual, though not the universally invariable, Australian manner. The Tasmanian dialects expressed neither gender nor number by inflection or agglutination, a remark which is generally true of the Australian dialects. The Tasmanian numerals were limited to one, two, three, four and five in the most copious dialects; the terms of the mainland were often enough limited to one and two. By which is not implied that the particular tribes could count no higher than the number of their highest numerical term, but that numbers above that term were expressed by combinations of the lower terms. Among the Tasmanians some tribes had numerals for one and two only, in which case the numerical system must necessarily have been binary, but others had distinct terms up to five inclusive. One form, however, which is given as the equivalent of five, seems to be a repetition of the term for *one* along with the term for *four;* this is the view which Dr. F. Müller takes of 'puggana marah,' and which I am disposed to take, but it is not affirmable absolutely.

The Tasmanians had such expressions as *legs-long* for *tall*, and they characterised certain affections, whether of the body or mind—*e.g.*, fear, hunger, fondness, &c., by names which indicated their effect upon the stomach or the eyes: features also of Australian speech. Another character common to both languages is the evident relationship of the terms for *cat, stomach, excrement,* and *ground*, the names for which appear below. It is common in Australian dialects to find the same word applied to *head* and *hill;* it seems to me that there was in one or two Tasmanian dialects the same idiom. The Tasmanians used diminutives, as for instance, 'pugga,' *a man*, 'puggetta,' *a child.* We are told by Mr. Curr* that diminutives were very common in the dialect

* Mr. Curr's "The Australian Race," vol. iii. p. 569.

of the Bangerang (Victorian) tribe.* Reduplication was a feature of both languages, though more general in the Australian, perhaps owing to its occurrence in Malay as well as Papuan speech. In both Australian and Tasmanian it was sometimes used to form the intensive mood, like the Heb. 'q‘taltal' or p‘‘al‘al'; *cf.* Tasmanian 'telbeteleebea,' *to eat heartily*, from 'tughlee,' *to eat*, and Australian (Kabi, Queensland) 'yeleliman,' *to speak quickly*, from 'yeli,' *to shout;* from 'ya,' *to speak.*

There is a feature of Victorian dialects which should not pass unnoticed : in several of them the first and second personal prououns depart from the usual Australian (Dravidian) 'nan-nin' type, a circumstance which supports the presumption of a strong disturbing element having been at work in Victoria. When we come to compare particular vocables, we find certain ancient forms cropping up in places very widely apart in Australia, a few fossils of an older stratum continuing in one more recent. The prototype of the modern Tasmanian is undoubtedly the stratum of which they are surviving representatives. While we may pick up one specimen here and another there all over the continent, just as in the case of other features we have noted, by far the largest number of words which are identical with Tasmanian forms or incontestable variants of them are to be found in Victorian dialects.

In the following comparisons the English word is usually the exact equivalent of the Papuan, but sometimes it is the general idea, at other times the etymological idea of the root, in which cases the particular meanings of the Papuan (Australian and Tasmanian) words are expressed. These analogies show how remarkably the old language protrudes through the modern Australian, like the primary rocks in mountain regions, piercing through the aqueous formations. The first table exhibits Tasmanian words, which are widely diffused in Australia, some of them appearing in places at great distances apart, and being unknown in the intervening space.

* I find that the Australian dialect most like Tasmanian in terminations is that spoken about the junction of the Darling and the Murray. The verbal terminations are in this case practically identical.

TABLE I.

ENGLISH.	TASMANIAN.	AUSTRALIAN.
Ground or earth	gunta, gonta, coantana (This may correspond to the Australian gunna or gudna, the common word for excrement)	nguntha (De Grey River); kuntha, *grass* (Cooper's Creek); thagound (Barnawatha); dagoon (Wellington); dha or tya (Australia generally).
Excrement	tiamena, tiannah, tyaner	nguntha (De Grey River); gunda (Shark's Bay); dagga (Bogan); duggan (Warren); thugga (Waljeers); gunang, gunna, or gudna (Australia generally).
Foot	pere*	mamberie (Lachlan and Murrumbidgee Junction); piru, *thigh* (Hopkins River, Victoria); pur-ring, *knee* (Lake Tyers, Victoria); piri, birri, birring, bret, occur as *nail, finger, hand,* or *footprint.*
Eat	tuwie, dodani, tuggana, tegurner	dha or tya (general).
Mouth	kakanninah, kaneina, canina, canea	gaad, kaat, kanek, korn, cone (all Victorian); knine (Waljeers); ngang (Deniliquin). These may be regarded as all embraced in one Papuan region—viz., the Victorian.
Fire	une, wighena, winnaleah, weenah, *wood*	wi, win, wee, ween (general).
You	neener, neena	indu, ngindu (general). Introduced here, but the writer regards this word, in its usual Australian form, to be Dravidian.

* The introduction of this analogy is justified by the following considerations. In the Tasmanian Vocabulary compiled by M. H. de Charency, 'pere' is given as a word for foot used in the south-east of the island. Along with which are given 'perelia' and 'pereloki,' *toe-nails.* The words 'perring,' 'paring,' and variants signify *footmark* in west and north-west of Victoria. In sandstone caves (Mr. Curr's "The Australian Race," vol. ii. p. 476), on the Cape River in Queensland, there are red impressions of hands which the blacks call 'beera,' although the local word for *hand* is 'buka.' The word 'beera' (or words almost identical) still signifies *hand* or *fingers* in the south of Queensland, on the Bogan River in New South Wales, and in Gippsland, Victoria. As the same word is used in places for *breasts,* I am inclined to think that the root, 'bir' or 'pir' originally meant any protuberance or extremity, and became specialised for such members as hands, feet, toes, fingers, &c. A Tasmanian term for *two* given as 'bura,' 'boula' or 'pooalih' is sometimes compared with Australian 'boolla,' *two.* I have omitted this analogy as doubtful. Several Tasmanian negatives are represented in Australia, *e.g.,* Tasmania 'parragara,' 'pothyack,' have analogues in Victorian 'boraka,' 'barapa,' New South Wales 'barre,' Queensland 'bar,' also in Queensland 'kurra.' Tasmania 'mallya' is represented by Victorian 'ngalanya,' South Australian 'madla,' West Australian 'marla.'

TABLE I.—*continued.*

ENGLISH.	TASMANIAN.	AUSTRALIAN.
Smoke	prooana, boorana	boort (Victoria); pooya (Streaky Bay); bwoya, boyer (Western Australia); pooyoo (common).
Tongue	tullana	tallan, tyelling (general).
Nose	mudena, minarara, moonar, manewurrar, mununa, muggenah, muye, muanoigh (The common element is ' mu ')	muntyin (Princess Charlotte Bay); moolya, moodla, moolla, mooroo (all common). The persistent element also ' mu.'
Thigh	tula, trungermarteener, teigna	dhirang (general).
Walk	yange (in Milligan's Dialogues)	yango, yanga, yan (general).
Speak	oona (in Milligan's Dialogues); oana, oanganah, *inform, tell;* oghnemipe, *to answer;* oghnamilee, *to ask*	wangow (Swan River); wangondi (S. Australia)
Water	mookaria, moga, moka, mocha, mookenner; moonghenar, *urine,* mungana, *urine* I would ask special attention to this analogy as being perhaps the most remarkable of them all. Just imagine two peoples perfectly separated for many centuries and without writing, yet retaining a word of four syllables in forms so precisely alike that even foreigners independently can spell the word 'mookaria' and 'muckaria'	mokkera (Murray and Darnug Junction); mukkara (Torrowotto Lake); muckaria and mugair (Lachlan and Murrumbidgee Junction); moogabaa (Alice River); mookorar (Port Macquarie); maicheri (Piangil); mittnk (Lake Boga, Victoria). All these Australian words are terms for *rain,* but their identity with the Tasmanian analogues is perfect, and terms for rain and water respectively are often interchanged among Australian dialects.
Bosom or Breast	paruggana (*a woman's*), parrungyenah (*a man's*). To these may be added the first syllable of proogwallah, a word meaning *milk,* and probably literally *breast-water;* if so, 'wallah' has an equivalent in Australian 'walla' or 'wolla,' common in New South Wales for *water* or *rain*	birri, *man's breasts* (Kamilroi Dialects); birring (Healesville, Vict.); brim brim (Mordiyallock, Vic.); birrin (Warren, N.S.W.); beergin (Forbes and the Levels, N.S.W.).

TABLE I.—*continued.*

ENGLISH.	TASMANIAN.	AUSTRALIAN.
Fly .	monga, mongana, moun-ga, *flyblow*, also *to buzz*	mookine, mugguing (on Culgoa River, N.S.W.); moongin, muggin, mogan, &c., forms for *mosquito* in Kamilroi ; mungi, *mosquito* (Piangil) ; moaing-moaing, *mosquito* (Kulkyne); mianong, myanga, *fly* (about Port Jackson); kerramongera, *fly* (Gippsland) ; moneya, *mosquito* (Omeo) ; miangan, *fly* (Omeo).
Come . .	tutta watta, todawadda (*come here*)	waarta (Lake Hindmarsh, Victoria); woti (Talbot, Victoria) ; ouarto (S. Australia); now-wunty, *come on* (Gippsland); wotte, wurte, *come on* (The Glenelg, Victoria) ; kakawattake, *come on* (Hopkins River, Victoria); kowatha, *come on* (Cooper's Creek); wat, watto, *away, off* (W. Australia).
Shout . .	palla-kanna (kanna means *to make a noise*)	curn-deeo (Mount Talbot, Victoria) ; kurnda (Swan Hill, Vic.) ; kanyandiga—ganyanda, *to call* (Lower Goulburn, Vict.) ; garnda, coerndee (Lake Hindmarsh, Vic.) ; kinda, *to call* (Mount Rouse, Victoria).
Eye . .	namer-eca, nam-muruck, nubreah, mongtena, moygia	mir (common Australian form); numuru (Daly River, Northern Territory). For Victorian forms, see Table II.
Arm . .	wornena, wu'hnna, gouna, houana ; wayeninnah, *elbow*	wooruk (Mount Rouse, Vict.) ; wing (Upper Murray, Victoria) ; wunyea (Lower Murray, Victoria, also on Darling and Murrumbidgee); wurt (Lake Hindmarsh, Vic.); whoornang, *forearm* (Lake Tyers, Vict.) ; wornick, *forearm* (Maryborough, Victoria) ; oona (Mount Freeling to Pirigundi Lake, S. Australia).
Sun . .	loina, loyna	laong, laank, *eye* (between the Lachlan, Murray and Darling, N.S.W.) ; arlunya, *sun* (Alice Springs, Telegraph Station, S. Australia); allunga, *sun* (Charlotte Waters Telegraph Station, S. Australia).

In presenting the Victorian-Tasmanian analogies, which are very numerous, embracing nearly all the words of the preceding table and including many more, I place in the front those having ' l ' as the initial letter.   This was the class of analogical words which first arrested my attention.   As initial ' l ' was a notable feature of the Victorian dialects distinguishing them from those of New South Wales and Queensland, and was also a peculiar feature of the Tasmanian language, I surmised that a common lineage was the reason for this likeness.   After comparing all the Victorian words beginning with ' l ' obtainable by me with Tasmanian words having the same initial, I find so large a number in the one set, evidently identical with words in the other as to be very surprising, especially when we think of the length of time which must have elapsed since the lines of language divaricated.   I have therefore come to the conclusion that Victorian words with ' l ' initial are lineal descendants of the primitive Papuan.

It is one of the recognised tests of the truth of a hypothesis that it opens the door to facts other than what was first discovered by it.   This test can be applied to verify the hypothesis here enunciated regarding this class of words having initial ' l.' By its means I have discovered that at least in Australia, and perhaps in Tasmania, ' l ' and English consonantal ' y ' have been at one time confused and perhaps coalescing and interchangeable sounds.

Professor Max Müller gives* some instances of the "confusion between two consonants in the same dialect," which he regards as a characteristic of the lower stage of human speech.   There seems to have been a very ancient confusion of this kind between the powers of ' l ' and consonantal ' y ' in Australia.   English ' y ' or ' i,' when consonantal, may very easily, through defects of hearing or utterance, be confused with ' l,' and the two sounds are to the ear closely related.   Examples are common enough. Compare the Indian corruption of ' Les Anglais ' to ' Yankee,' such forms as Italian ' piacere ' for Latin ' placere,' and the French pronunciation of such combinations as ' eille.'   That in Australo-Papuan speech ' l ' and consonantal ' i ' have been confused the following examples will make sufficiently clear.   In western and north-western Victoria such forms as these prevail

* Max Müller's " Lectures on the Science of Language," vol. ii. pp. 188-89.

as equivalents of *black woman*: 'Leyoorook,' 'leurook,' 'liarook,' 'leyoor,' 'lioo,'[*] which may be compared with 'yewa' (Cooper's Creek), 'yooratoo' (Unyamootha Tribe), and the following forms found in Western Australia from Perth southwards: 'Yokka,' 'yooko,' 'yawk,' 'yorka,' 'york.' The next example is a word which throughout the greater part of Victoria begins with an 'l.' It is the term for teeth, and the following are typical Victorian forms: 'Lianyook,' 'lea,' 'liia,' 'leor,' 'leurn,' with which compare such forms as 'yira,' 'ira,' 'eera,' 'yeera,' very widely distributed throughout Australia, except in Victoria. It is interesting to note that, with over a thousand miles of country intervening in which the 'y' prevails, the Victorian and Tasmanian type in 'l' appears again at Caledon Bay, on the Gulf of Carpentaria, in the word 'lerra,' indicating the persistence of a Papuan pronunciation due to a backward and northward eddying of the Papuan speech.

Then further, as the word for *stone* we have in Victoria 'larr,' 'laa,' 'la,' while on the Murray the same word is represented by 'yarnda,' on the Lower Bulloo River and at Yelta by 'yernda,' and in the Woolna language again, in the far northwest, the form 'lunga' occurs. A similar word in some places is used for *camp*, assuming such forms as 'larrh,' 'larrer,' 'lerra,' which even in Victoria is represented by 'ira' at Lake Hindmarsh, 'iray' at Tatiarra, and by 'ieera' at the Gawler Range in South Australia. The same interchange of 'l' with 'y' is observable in certain terms employed to designate *skin*, *bark*, and *canoe*; 'look' and 'looko' are words for *skin*; 'long' and 'laikoti' for *bark*; 'longoi' and 'longwe' for *canoe*, with which compare 'yangoibi,' 'yongoe,' 'yoongoip,' 'yungoot,' all meaning canoe. It should be observed that by a radical or natural metonymy *skin* and *bark*, *bark* and *canoe*, are frequently expressed by the same word. The original unity or early confusion of the letters 'l' and 'y' is illustrated by several words meaning thigh—*e.g.*, 'langui' at the mouth of the Leichhardt River, 'lar' at the mouth of Norman River, 'yungurra' at Porter's Range, 'yangara' at Upper Flinders River,

* I have to point out that the initial syllable in the Victorian words corresponds to a New Hebridean word for woman—viz., 'lai,' 'lei,' 'le '—also to a Tasmanian word 'lowa,' hence said first syllable may not be represented at all in the other Australian words compared, parallels to which are found in compounds in Victoria in the form '*-goork.*'

'yungera' at Cape River. The most interesting of the above
examples are the words for *blackwoman, stone,* and *teeth,* which
serve as a kind of bridge for crossing to Tasmania, or as links
to unite a particular class of words there with their continental
variants, and to widen the field of comparison while they ensure
the validity of the operation.

I shall begin the comparison of Tasmanian and Victorian
words with the particular class which first suggested their relationship to me—the words with initial '1.' Why should this
class of words be a phonological peculiarity marking a group of
dialects in south-eastern Australia, spoken in a tract of which
the northern boundary almost coincides with the Murray? Why
should this group of dialects be hedged round landward by
others distinguished by the absence of this very peculiarity?
Why should words of this phonic character exist plentifully
in Victoria and be comparatively rare in most other parts
of Australia, save in the extreme north-west and about the Gulf
of Carpentaria? Why in Victoria, and be also a pronounced
feature of the Tasmanian tongue? Why, I ask, unless there
linger in Victoria evidences of the most recent Papuan influence as compared with other parts of Australia, and sure
proof of the Tasmanians having had a closer affinity to the
Victorians than to the rest of the Australian natives?

It might at first sight be doubted whether the common
Victorian word 'layarook' or 'lyarook' and other variants is
the same as the general Tasmanian word 'lowa,' both sets of
words meaning *woman* or *black woman;* but fortunately the
Victorian word has retained the form 'laua' in Gippsland,
which, being phonetically identical with the Tasmanian word,
establishes beyond the possibility of a doubt the fact that the
words for *woman* (sometimes *wife*) in both languages are the
very same. This analogy is subject to the qualification that
the longer form has another element added. Without dwelling
further on particular words, I append a list for comparison,
and in order to indicate as near as may be known the root
form and to show the direction of divergence I give a
number of variants from both sides of the Strait. The
English word, as before, represents the general or etymological idea of the root, being also usually the exact equivalent of the native word in both columns, and the native words,

with a few exceptions, will be found either in Mr. Curr's or Mr. Brough Smyth's work.

TABLE II.

ENGLISH.	TASMANIAN.	VICTORIAN.
Woman . .	lowa, loa, loalla, loubra	lio, laua, *wife* (Gippsland); leyoor, leirock, layarook; loangko, *a wife* (Lower Murray).
Teeth .	leeaner, yanna; leeanner, *to bite*	lia, lear, leeunger, leanook.
Stone . .	longa, lonna, loine, larnar	long, *a cliff*, lang, lak, laugh, lar, laa; *cf.* also wollong or wallung, common in New South Wales. ['Lung' is a common word for stone in Central India, whatever inferences the fact may involve.]
Open or cut . [*The members of this group of words are almost certainly derived from one or other of the two preceding groups*]	leeang wellerary, leearway, laini *to untie;* lowgoone, *to cut;* larre, *to scratch;* lowoone, *to scarify;* lergara, leawarina, *to flay;* liellowullingana, *crevice* or *fissure*	loong gonak, *to divide;* larlgroo-war, lal-go-mak, *to split;* Lal Lal, probably *Great Rift*. [With this class of words compare 'lalingander,' *axe*, a word used by the Woolna Tribe, near Port Darwin.]
Home, house, nest, camp	line, lenna, liena, liee	laangy, langi, lar, larr, larnoo, lingi.
To sleep . .	loagna, logurner, lony	loomai (Woodford); loomia (Dartmoor); *cf.* also yooanan.
Tree, stick .	loyke, loatta; lottah, *gum tree;* lerga and lerina, *waddy*	lang; loang, *little tree;* lordwill, lurt, and lead, *stick*.
Top or point .	lyetta, *sharp* or *peaked;* letteene, *a peak*	lit, *point;* littia, *sharp;* litwong, *to sharpen*.
Serpent . .	loieua, louinabe, loina; lollah, *earthworm*	loowa birri, *wood snake* or *constrictor*.
Leg . . .	luggra, leurina, lurerener, langaner; 'lure, *ankle;* lugh, *foot*	lourko, lourt-am-nook; lourk, *calf of leg*.
Coal, charcoal .	loarra, loira	lourn (Gippsland).
Water . .	lia, leena, legana, lerui, line; liapota, *creek;* loyuleena, *spring;* lyaleetea, *sea*	larra, lajeranyen, *spring;* loortowi, loortokal, *creek;* Leaghurr, Lalanguite (names of Lakes); ludht, lowtoohk, *river;* lamat, *sea;* lakulang, *salt lake; cf.* also yallook, a common word for creek.
Child . .	leewoon, looweinna, *children;* luena, leuna, ludawinna, *boy;* ludining, *girl*	lathe, leed, *boy;* leech, *son;* lunden, lunduk, landhagonert, latingata, *sister*.
Big . . .	proina, proingba, paroina langtha, lackrana, *great*	porin, parok, parronk leengil.
Man, black man	pugga, pah, penna (*Capt. Cook*) beah	baang, peang (Central Victoria).

TABLE II.—*continued.*

ENGLISH.	TASMANIAN.	VICTORIAN.
Kangaroo .	terror, tarrana, tarrleah	tirrar, tyirra, jirrah (all in Gippsland).
Mother . .	pawamena, pamena, par-meny	bawain, parbine, baabin, par-buk, paab, paapa, papay, papi.
Opossum . .	wollimerner	wollert, wolard.
Head . .	poyta, poiete	poibi, poko, pooruk, pork.
Mouth . .	kaneina, canina, canea	kanek, koorn, gaat, cone.
Lips . .	wurlerminner	werrong, wuro, woortogno, wooro (common for mouth).
Eye . . .	namer-eca, nam-mur-uck, mongtena, moy-gta (initial syllable 'na,' probably con-nected with Austra-lian root 'na,' to see)	mirrenyook, mirnook, mir, mynook, mingi, myng. mooeh, mirnik [termination 'ook' may be sign of pos-session].
Ear .	wayee, wegge	wooring, wring, wing, weinye-duck.
Thigh . .	(*a*) tula, teigna, trunger-marteener ; (*b*) kaar-werra	(*a*) djereng, dering ; (*b*) kaar, karingatuk, karnook, kaar-chuk, karip.
Foot . .	tyentiah, teeantibe, *to trample—i.e., to foot*	tyenna, tyain ; tey-yan. *foot* or *footmark* (Gippsland).
Blood .	coccah	cookyangerack, gooak, koor-kook, krook.
Fire . .	une, onane, wighena, winnaleah	wee, wein, weeing ; wying, *light* (prevalent also out of Victoria).
Wood . .	gui, weenar ; weegeena, *deadwood ;* weena, *tree;* wiena, winna, *firewood*	ween, we, wing [presumably *firewood*].
Smoke . .	boorana, prooana, pro-goona ; boora, *rain*	boort, booring, poorin ; boo-rang, *fog ;* boorrarrang, *mist*
Moon . .	weeetta, weethae, vena, weena ; tooweenyer, *sun* or *moon*	huera, wana, waingmil, wyng-wil, waing, wyrng, *light ;* ngiwen, noween, noweyo, *sun*
Grass . .	poene	poon (Omeo) ; booite, boott.
Knife or flint .	teeroona, trawoota	teer, deer, tirr, taree, all mean-ing *tomahawk*.
To eat . .	tegurner, tuggana, tuwie	tunganeit, tukkali, thangarth, thaange, thana [most likely co-derivatives with 'dhang-ga,' a common Australian word for teeth].
To go . .	tawe ; tangara, *go on ;* tagara, *go away*	toewangeit ['·eit,' verbal ter-mination], tanna toa, *go away ;* tanna, go, in phrase 'tanna noul ?' *will you go with me ?*
To walk . .	yange, *they walk* (in phrase 'yange me-naye,' they walk along the river. In Milli-gan's Dialogues)	yangan, yannonan, yanga, yanna.
Stomach . .	ploner, plaangner	polloin, belanyin, ballingek, boole, beleni, belangee (Lachlan and Murrumbidgee Junction).

A few of the foregoing analogies may not seem to be sufficiently established, but even if 10 per cent. were to be discredited for want of certainty, a great array of obvious ones would still remain to attest the kinship of the two languages. A brief notice of two or three more Tasmanian words may be helpful to relate the two forms of speech. In Tasmania there are two general but widely dissimilar classes of appellations for the kangaroo, one represented by such terms as 'lyenna,' 'lathakar,' 'leigh, 'lurgu,' the other by 'terrar,' 'tarrana,' 'tarr.' The former class seems to be connected with one of the Tasmanian sets of terms for *leg* or *foot*, exemplified by such words as 'luggana,' 'lugh,' 'leoonya,' 'luggra,' 'lathanama.' I have been unable to discover any similar word for kangaroo on the mainland except 'langootpa' at Port Darwin, and 'loityo' at Caledon Bay, on the Gulf of Carpentaria. The one set of words for kangaroo being almost identical with the terms designating the organs of locomotion, it seems pretty certain that either from superior power in the use of these or for the long measure of them with which the kangaroo is endowed, the animal received its name.* And it seems at least probable that the 'l' words for *leg* and *kangaroo* in Tasmania are related to the universal Australian word 'yanna,' or 'yango,' *to walk*, for which the form 'lingo' is said to occur in Victoria.† It has already been shown that both in Australian and Tasmanian speech 'l' and 'y' when initial have often been confused.

The words 'locko,' *foot*, in use at Caledon Bay; 'langiu,' *thigh*, used at the mouth of the Leichhardt River, and 'lar,' *thigh*, at the mouth of the Norman, on the Gulf of Carpentaria, are curious relics of the original terms for either the motors or the motion, and reach without other connection right across the continent southwards to claim kin with their Tasmanian friends having initial 'l.' This line of argument is powerfully corroborated by the occurrence of the following terms for *thigh* in Queensland ; 'yungurra' at Porter's Range and Walsh River ; 'yangara' at Upper Flinders River ; 'yungera' at Cape River

---

* I derive the word 'kangaroo' (originally spelt '*kanguru*') from 'ka,' *nose*, or *head*, and 'gura,' *long*.

† Mr. R. Brough Smyth, "The Aborigines of Victoria," vol. ii. p. 127 ; phrase, "Where are you going ?"

means *lower part of leg*. At Menindie 'yango' means *left thigh*, and 'yalko' is the word for *thigh* on the lower Paroo, the Warrego, and at Weinteriga, on the Darling.   All these are evidently enough the local equivalents of 'langiu,' used at the mouth of the Flinders River. What if 'yan,' *to go*, be a related word? If. so, the term 'yungar' or 'youngar,' applied to *kangaroo* throughout the greater part of Western Australia, may be cognate. Perhaps it may be too fanciful to pass northwards to Java and connect the foregoing words with Javanese 'laku,' 'lunga,' 'lingar,' *to go*, but the likeness in sound and meaning is very tempting.

The affinity subsisting both in Australian and Tasmanian between the terms for *ordure, intestines*, and *ground*, cannot fail to force itself upon the investigator's notice. The words are given in the tables above, and when compared they lead to the conclusion that they are inter-related in both regions, the common element being 'ta' or 'tia.' Alongside of these may be placed the radical part of the words meaning *to eat*: 'tegurner,' 'tuggana,' 'tuwie' (Tasmanian); 'tunganeit,' 'thaange,' 'dha,' 'dhoman' (Australian); and perhaps the very prevalent Australian word for *teeth*, 'dhangga,' with variants. This last word may be onomatopœic and the other vocalics may narrow down to two roots, perhaps only related in likeness, one meaning *to eat*, the other *ground*. Another class of related words in both languages, and forming all together a group of related words, are the equivalents of sun, moon, light, fire, eye. In both languages the radicals 'na,' 'mir,' 'wi,' the etymological ideas of which are respectively *see, eye, fire*, are combined in numerous ways, making such compounds as 'see-fire' (*sun*), 'see-eye' (*eye*), 'fire-eye' (*moon*), and the like. With these may be compared a second set of words for *eye* and *sun*, taking the form of 'loina,' 'lunya,' and the like; also occurring in the language of both peoples.

Mr. Hyde Clarke has endeavoured to show* the affinity of the Yarra dialect with dialects of Mozambique, and later Mr. E. M. Curr essays to prove the kinship of African languages generally with those of Australia and Tasmania. Physical characteristics

* The Yarra Dialect and the Languages of Australia in connection with those of Mozambique and Portuguese Africa ("Transactions of the Royal Society of Victoria," vol. xvi.).

alone would suffice to obtain acceptance for Mr. Huxley's view that of all races the Papuan is most nearly related to the African. And besides physiological considerations, certain practices and superstitions common to the Australians, Tasmanians, and Africans point to identity of ancestry at some far distant past date, but the verbal analogies adduced are rather shaky props on which to rest the relationship argument. Mr. Clarke avowedly discards, as Mr. Curr does tacitly, the testimony from grammatical structure, and they both present merely phonetic resemblances, which may be very misleading, as the following considerations will show.

In the first place, there is a number of vocables which may be looked upon as universal; whether they be of onomatopœic origin or no, does not affect the present argument, but the words are as much European or Asiatic as they are African and Australian. They occur as equivalents for father, mother, breasts, milk, teeth, tongue, eat, go, and are such roots as 'ba,' 'pa,' 'ma,' 'ta,' 'yo.' Further, the possibilities of speech are limited; all races have virtually the same vocal instrument, and there is, I believe, in mankind generally an inherent capacity to name things according to the subjective effect which the observation of them produces, giving good grounds for recognising the ding-dong theory as partially (and in large part) accounting for the origin of language. And therefore, if phonetic likeness alone were to be taken into account, a very good case could be made out for the descent of the Australian speech from the English or *vice versâ*, especially if, when the English dictionary failed, we might call in the aid of any language on the continent of Europe to supply the deficiency; and indeed, for such comparison, all the Indo-European languages might be regarded as one family.

With some two hundred dialects to draw upon in Australia, and dialects innumerable in Africa, it would be strange indeed, the possibilities of human speech being limited, if close coincidental resemblances were not discoverable here and there in the compared regions among appellatives for the same object. There seem to be no solid reasons for deriving the Tasmanians and Australians independently from the Africans, if it be right to say that they are sprung from the Africans at all. It is perhaps nearer the truth to say that the Tasmanians, in common with

other Papuans, are of the same stock as the Negroes, the common
ancestry being neither Papuan nor Negro, or as much the one
as the other, and that the Australians are derived from the same
original stock through the Papuans with a strong foreign admixture.

Latham having suggested New Caledonia as the probable
temporary home of the Tasmanians on their way to their last
resting-place, it will be well to inquire here what grounds there
may be for falling in with the suggestion. In physical
appearance, and especially in complexion and quality of hair,
the aborigines of New Caledonia, like other dark Papuans, bear
a strong likeness to the Tasmanians. There is no better basis
for Mr. Latham's suggestion beyond this likeness and the
surmise that, as it seemed improbable that Tasmania had been
peopled from Australia, its inhabitants might possibly have
drifted from the nearest settlement of Papuans most resembling
themselves in appearance. Of the New Caledonian language I
have only been able to see specimens given by Gabelentz in his
*Die Melanesischen Sprachen.** The phonic combinations resemble
more the Australian than the Tasmanian. The only words
which I can find that might be related to either Australian
or Tasmanian words indifferently are 'mainya,' 'mandig,'
'muanden,' 'muala,' *nose;* 'dendan,' *to come away;* and
'adheya,' *foot.* Certainly few and doubtful analogues. A
peculiarity of New Caledonian is the use of different forms
of numerals according as an object is animated or not. The
pronouns, in having a dual resemble those of Australia, and,
so far as can be known, differ from those of Tasmania.
Viellard mentions that '-ri' and '-ra' are suffixed to substances
to indicate *whose* and *which* respectively, a feature unknown in
the Australian and Tasmanian. There is no necessity for further
comparison. The conclusion from the only available evidence
is not in favour of affinity between New Caledonian and the
other two languages. Its phonetic system is smoother than
that of Victoria and Tasmania, but not so fluent and musical as

* I have since examined the lengthy New Caledonian Vocab. given in
"Vocabulary of Australian Dialects," printed for the Melbourne International
Exhibition of 1866, with the result of finding two or three more words that
might be related to Tasmanian equivalents, but no evidence of even so close
a relation between New Caledonian and Tasmanian as between the latter and
Australian.

that of central and northern Australia, and the data, instead of suggesting that Tasmanian is more nearly akin to New Caledonian than to the language of the mainland, favour the very opposite conclusion.

The writer ventures to affirm that future research will only tend to corroborate the opinion which he has here enunciated and endeavoured to establish—namely, that Tasmania was first peopled from the Victorian shores. The point from which the emigrants left the mainland was probably Wilson's Promontory, from which a string of islands runs like stepping-stones across the Strait, which were perhaps at one time larger and more numerous than they are now, if they did not form an isthmus. It does not follow, however, that the most distinct vestiges of the old Papuan Australians should be found at this point. From philological considerations it would rather appear that the Lower Murray, and perhaps the Lower Murrumbidgee, served for long as a natural defence to the Victorian Papuans, and that the invaders poured into Victoria across the Upper Murray, took possession of central Victoria, pressing those who were being dispossessed back on either flank. At all events, the most numerous and on the whole the clearest verbal analogies with Tasmanian are to be found in north-western Victoria from Lake Boga northwards, and about Bumbang, Tatiarra, and Piangil on the Murray. This markedly Papuan class of dialect extends on a line up the Murrumbidgee and embraces a large tract of country between this river and the Lachlan above their junction.

Having now demonstrated, beyond all question it is hoped, that the Tasmanians were the lineal descendants of the primitive Australian race, that the substratum of the modern Australians is Papuan of the same blood as the Tasmanians, and, as might naturally be expected, that the quarter of Australia which lies nearest to Tasmania retains most distinctly traces of the indigenes, the next duty is to attempt to disentangle and identify the other elements which go to constitute the Australian race as it now is.

CHAPTER III

THE DRAVIDIAN ELEMENT

UPON the original Papuan stock of Australia there must
have been grafted a very strong scion from another and in
some respects very different stem, and the union must have
been effected in the remote dim past, the stock from which the
graft came having since then altered by progressive development almost beyond identification. The people who formed this
fresh addition to the primitive race had probably a lighter complexion and straight hair. What impelled them thither we
know not. We are familiar with the idea of successive waves
of population starting from a common centre and being arrested
only by an uncrossable ocean. History and philology together
have related to us how Roman and Teuton followed Kelt until
the broad Atlantic stayed their occidental march. A Semitic
population pursued the sons of Ham bearing the ancestral curse
of servitude into the utmost recesses of the dark continent. It
is left on record, both in parchment and in temple ruins, how
the Buddhists were driven out of India in the seventh century
of our era, and had they not found congenial soil in Java they
might have continued their southward course and left their
mark on Australia. But the fact that they came so near to the
southern continent is an indication perhaps of the track of the
line of least resistance to a fugitive people; at all events, their
migration hints at the channel along which might have flowed
former streams of humanity expelled from India or its neighbourhood by irresistible pressure from the north.

Although the Australians are still in a state of savagery and
the Dravidians of India have been for many ages a people

civilised in a great measure and possessed of a literature, the two peoples are affiliated by deeply marked characteristics in their social system and by sure affinities in language.

A most striking peculiarity of the Australian system of kinship had been recognised and published long before the late Rev. W. Ridley stated it and carefully traced it out, but to him is due the honour of accurately formulating its details (as it exists among the Kamilroi), and the Rev. Dr. L. Fison is to be credited with having clearly established its identity in essentials with the Tamulic system. As Australian marriage and consanguinity will be treated subsequently, it is needless to do more here than state that in certain important particulars Dr. Fison, with the aid of Mr. Ridley, has demonstrated the identity of the Dravidian and Australian systems of kin. The sum of these particulars is contained in the following proposition, which is equally true for both peoples, and holds in it the root principle of the system of kin: "A being a male, his brother's children are considered his own children, his sister's children are his nephews and nieces; his sister's grandchildren as well as his brother's are considered his grandchildren."* Let A be a female, then with the interchange of the terms 'brother's' and 'sister's' the proposition is also true. The relational nomenclature is such as would arise if a group of brothers were joined in a communal marriage with a group of sisters. And further, "in Tamil the elder brother is distinguished from all the rest by the title brother,"† and the Australian practice indicates some similarity of thought to this.

If so strong a bond unites the aborigines of central and southern India with the majority of the Australian tribes, among the latter exceptional departures from the prevailing type of relationship nomenclature cannot invalidate the conclusion as to its source.

Besides the powerful token of affinity to aborigines of Hindostan supplied by the possession of the same social groundwork, Australia bears also linguistic marks of Indian connection so deeply and widely impressed as to be indelible, and to serve as one of the most powerful and conspicuous bonds of union among the Australian dialects.

* Rev. W. Ridley's "Kamilroi and other Australian Languages," pp. 164. *et seq.*   † *Ibid.*

First of these linguistic marks may be mentioned the syllabation preferred by the genius of the Australian tongue. Like the Dravidian, it is extremely simple and averse to compound or concurrent consonants. In Tamil,* " double or treble consonants at the beginning of syllables like 'str' in 'strength' are altogether inadmissible. At the beginning, not only of the first syllable of every word but also of every succeeding syllable, only one consonant is allowed. At the conclusion of a word double and treble consonants like 'gth' in 'strength' are as inadmissible as at the beginning, and every word must terminate either in a vowel or in a single semi-vowel, as 'l' or 'r,' or in a single nasal, as 'n' or 'm.' These observations are just as true of all the dialects in Australia save those of the south-eastern and south-western corners, where the softer syllabation has been unable to displace the older harsher Papuan.

The next point of contact to be noted is the agreement of the stems of the Australian first and second personal pronouns singular with the Dravidian. Mr. Norris is said to have been the first to point this out, but on comparison the conclusion is inevitable to the most casual observer, the fact being self-demonstrative. Logan says that the roots of the Dravidian pronouns are 'na,' 'en,' 'ne,' 'an,' *I*, and 'ni,' *thou*. Speaking generally, these are the persistent stems of the same pronouns throughout Australia, the prevailing forms being 'ngai' or 'ngan,' first person, 'in,' 'yin,' or 'ngin,' second person. In Victoria, again, there are the greatest and most numerous divergences from the typical forms, evidencing the more recent clash with another speech.

Caldwell notes that the Telugu forms its pronominal plurals by prefixing 'lu' to the singular, and compares this with the Australian additions 'lu,' 'li,' 'dlu,' 'dli,' &c., employed for a similar purpose. He also adduces a more Australian-like instance—viz., the Dhimal, on the north-east frontier of India, which has 'nâ,' *thou;* 'nyel,' *you*. The same writer further suggests a likeness between Tamil accusative 'ennei,' *me*, and the Australian 'emmo,' *me;* but a much better comparison may be made between 'ennei' and 'nganna,' the common

* Caldwell's "Dravidian Grammar," p. 138.

Australian form for *me*. Other verbal roots common to both classes of languages might be cited, but nothing special could be adduced from them, inasmuch as they are not peculiar to these two classes. It is different with the pronominal stems just considered, for in both cases they are distinguishing features, and it is a very natural inference that the language which teaches a nation to say 'I' and 'thou' must be one of its very early and most influential pedagogues. Mr. Caldwell further shows agreement between Dravidian and Australian in the following particulars*: The use of post-positions (a feature, however, on which stress should not be laid, as it was very pronounced in Tasmanian); the use of two forms of the first person plural, one inclusive of the party addressed, the other exclusive (a feature also of South Sea Island languages); the formation of inceptive causative and reflexive verbs by the addition of certain syllables to the root, and generally, the agglutinative structure of words and the position of words in the sentence.

There are other very marked resemblances of which Mr. Caldwell was unaware. In Telugu, 'yokka' or 'yoka' is sometimes appended to the inflection or natural genitive as an auxiliary suffix of case—*e.g.*, from the ordinary possessive 'na,' *my*, is formed optionally the equivalent form 'na-yokka,' *my*, *of me*, with which may be compared 'nganyunggai,' *my*, in the Kabi (Queensland) dialect, and various forms in '-yuck' occurring in Victoria and elsewhere.

Caldwell† calls attention to a marked divergence between Telugu and Tamil in their respective terms for *one*, which are *oka* and *oru*. He infers that there existed an original basis of both of the form *okor*, like Samoiedian *okur*. A similar etymon—namely, *kuru*, *one*, varied to *kula*, *uru*, *gura*, *koo-took*, &c., with often the affix *pa* or *po*—covers nearly the whole of Australia. It is the regular term for *one*, introduced by the second tide of immigration. Another form in Tamil is *oruvan*, *unus*. Canarese has *obbanu* = *or-b-an*. They are quite like the Australian.

The Dravidian languages‡ are destitute of any common term

---

* Caldwell's "Dravidian Grammar," p. 53.

† Caldwell's "Comp. Gram. of the Dravidian Languages," p. 243.

‡ Caldwell's "Dravidian Grammar," p. 477.

for brother, sister, aunt, &c., and use instead a set of terms which combine the idea of relationship with that of age—*e.g.*, elder brother, younger brother, and so forth. This applies generally to Australian speech. "In the Dravidian languages the second person singular imperative is generally identical with the root or theme of the verb; this is so frequently the case that it may be regarded as a characteristic rule of the language."[*] The same may be said of some at least of the Australian dialects. Compare Dravidian 'varu,' *to come*, imperative 'va,' with Kabi (Queensland) 'baman,' *to come*, imperative 'ba.' Several years ago I wrote of the verb in this dialect: "The simplest part is the imperative, which commonly consists of one syllable and very rarely exceeds two." "It is a remarkable feature of the Dravidian languages that they have no relative pronoun whatever."[†] This is also a feature of Australian speech. "The mode in which a language forms its preterite constitutes one of the most distinctive features in its grammatical character, and one which materially contributes to the determination of its relationship."[‡] Tamil forms its preterite by adding 'd,' which for euphony is sometimes preceded by 'n,' "owing to the Tamil fondness for nasalisation," says Caldwell. This may or may not be the reason for the appearance of the 'n,' but the common form of the preterite in Kabi, Wiradhuri, and other Australian dialects terminates in 'n.' In the Dravidian the accent is on the first syllable. This is commonly the case in Australian, and is easily accounted for by the agglutinating character of both languages.

It is a formidable objection to the theory of the relationship of Dravidian and Australian speech, that so distinguished a philologist as Dr. F. Müller, who was on the scientific staff of the *Novara*, should have declared emphatically against it. He says that, viewed even apart from the racial difference, the glossarial affinities are too weak to support the affirmation that the languages are genealogically related. There are, he adds, certain points observable which lead to the conclusion that such connection is impossible (*unmöglich*). Now for his argument. He asserts that if a genealogical relationship existed, it would receive fullest expression in the speech of

[*] Caldwell's "Dravidian Grammar," p. 420.

[†] *Ibid.* p. 412.   [‡] *Ibid.* p. 390.

West Australia, which is geographically nearest the Dravidian languages. But this is an unwarranted assertion, based upon the assumption that affinity of speech depends upon proximity of residence in a bee-line. Whereas it is, I hope, clearly proven in this essay, that migration was from the north-east, not from the west, and that the west was one of the corners into which the purer Papuan race was forced. Further, he affirms that the ' nan-nin ' type of pronoun prevails more or less in Thibet, China, and elsewhere, as well as in central India. A good argument, but the likeness is not generally so close. He further objects to the rules of class-marriage being introduced as evidence of relationship, because similar regulations are found in other parts. I think, however, that the likeness between those of India and Australia is most marked.

The last resemblance that I shall mention is the occurrence in both Dravidian* and Australian languages of a negative imperative or prohibitive particle. For instance, in the Kabi dialect, most referred to because most familiar to the writer, with the imperative when prohibitive the word or particle ' bar ' is used preceding the verb; on all other occasions other negatives are employed. This is a feature of South Sea Island languages also.

If there were only one or two resemblances like those enumerated between the two classes of languages, they might be passed over as purely coincidental and not due to a common derivation, but the resemblances are too numerous and striking to be so lightly dealt with, and can only be referred to a strong family likeness. As more Australian data becomes accessible there is no doubt that an exhaustive comparison will well repay for the labour, and it may be found that Dravidian and Australian languages are mutually explanatory.

The famous Australian boomerang may be another means of establishing connection with India, where the weapon is also found, the kind which returns to the thrower being, however, so far as is known, confined to Australia. We search the Malay and Papuan armouries in vain for any trace of it, and are therefore obliged to credit some other race with its introduction to

* Caldwell's "Dravidian Grammar," p. 36.

Australia, unless we unnecessarily assume that it was invented here independently. The boomerang is used in Africa about the upper course of the Nile, but we need not travel so far for it across barriers that might be termed impassable when it is obtainable so much nearer and in a place from which, as we have seen, a highway has led thither almost to Australia's shores. If the framework of society and those terms which are almost as close to a man as his own name, have both been introduced from India or its neighbourhood, it requires no stretch of imagination to suppose that the boomerang came along with them.

The Australian religious superstitions point rather to a connection with the South Sea Islands than with India, or as much to the one as to the other. In each of the three regions there is veneration for smooth pebbles. This is evidently. a very ancient religious sentiment. Isaiah charged the Jews with this form of idolatry.* " Among the smooth stones of the stream," was their portion ; "even to them " had they " poured a drink offering " and " offered a meat offering." In India the worshippers of Vishnu venerate a kind of pebble called *Salagrama* ; specimens that have been seen by Europeans are said to range from the size of a musket-ball to the size of a pigeon's egg. The particular sorts have an aperture with four spiral grooves in the perforation. The Hindoos believe that these apertures are the traces of Vishnu having entered the stones in the form of a reptile. It is worthy of note that among the New Hebrideans, as the Rev. Dr. J. G. Paton has told me, the sacred pebbles have a small aperture, regarded as the place of exit and entrance for the spirit which the stone represents. The *Salagrama* stones are found in the bed of the Gundak River, and are supposed by Coleman to be mineralised fossils of the Belemnites or Orthoceratites. The *Binlang* stones found in the Nerbudda River are worshipped as emblems of Siva. The veneration, then, of smooth stones would seem to relate the Australians equally to the Hindoos and the Kanakas.

There may be another connecting link between the Dravidians and the Australians in the emblematic use of a red right hand daubed on rocks in various parts of Australia, generally about caves. Dr. Carroll, in an article contributed to the *Centennial*

---

* Isaiah, lvii. 6.

Magazine (October 1888), affirms a connection. He says the red hand "is still symbolic of the various attributes of Siwa, the Punisher, Avenger, or Destroyer of the Hindu." My examination into Indian mythology has failed to make this quite so clear, which I admit is only a negative argument, and therefore not entitled to the weight of a positive argument, unless the field of negation be exhausted. But I find that in figures of the goddess Maha Kali, a form of Parvati the consort of Siva, a number of red-palmed hands are delineated. There are seven red hands pointing downwards, forming a cincture about the waist. The functions of Maha Kali are variously explained. Human sacrifices are offered to her. She is said to represent the active energy of all-renewing time, but sometimes she personates time as destructive. It is therefore possible that the red hand blazoned on Australian rocks may relate the Australian to the Dravidian, but, as considerations to be brought forward later will suggest, the great probability is that this symbol was introduced, not by an Indian race direct, but by a Malay people, who have certainly carried hither sure tokens of Hindoo mythological influence, as will be demonstrated when Art and Religion are dealt with.

CHAPTER IV

THE MALAY ELEMENT

UNIVERSAL and strong though it be, the so-called Dravidian
influence is insufficient to account for the great divergence of
the Australians from the pure Papuans in physical features and
in language. Another cause must be posited, and is to be found
in the Malay element.   Since British colonisation the Malays are
known to have frequented the north and north-west coasts.
Mr. Curr is of opinion that their visits are only of recent date,
and quotes in support the statement of a Malay named Pobassoo,
whom Flinders met at the Gulf of Carpentaria in 1803. He
professed to have been one of the first of his countrymen to
visit Australia.* The historical knowledge of the Malay would
not penetrate many centuries backward, and moreover, his
evidence is overborne by the physique of the people, in the
north especially but elsewhere as well, by the naturalisation of
a number of important Malay words, such as the term for *teeth*,
a change which mere visitors could not effect; and further, by
faces of the Malay type and pure Malay words appearing in
localities far removed from casual intercourse, as for instance, in
the extreme east and west. There are several old camps of
Malay *bêche-de-mer* fishers on the north-west coast.   I am
informed by Mr. Bradshaw that at one of these, on a small
island near Osborn Island, Captain Hilliard saw some old tama-
rind-trees, introduced presumably by the Malays, and the age
of the oldest tree was apparently some two hundred years.

As in the case of most other races, there would have been—

* Curr's "The Australian Race," vol. L. p. 271, *note.*

indeed has been with the Malays—a time of special activity and expansion. Coming then to Australia, they would be unable to enter into commercial relations with its poverty-stricken, nomadic, naked people, and those who did not return to their own land would simply settle down to a life of indolence and sensuality and melt like snowflakes in a sea of human life. But a shipload of Malays attaching itself to an Australian community would not be absorbed without leaving some traces of its presence.

If the Malays, arriving in Australia even in twos and threes, did not set themselves deliberately to teach and elevate the people, but sank to the same savage level, they could not possibly disappear with their unconscious influence absolutely obliterated. This influence is especially noticeable in the physique of the people in the north. They are more slender than the rest of the Australians, have less hair on the body, and their skin is fairer. Sir George Grey, in one of his journeys, saw three men of a fair race resembling Malays, and some of his party saw a fourth.* This was near the cave where he discovered some paintings of clothed people. These four men might, no doubt, be the posterity of one or two castaways. But even in the south of Queensland I have seen several faces distinctly of the Malay type, with the nose snubby and rather small and the skin of a dark copper colour. Occasional instances of sullenness and taciturnity among the Australians are probably the result of Malay ancestry.

The admixture of Malay blood goes far to account for the difficulty, that a race such as the Australian with a Papuan basis should have the hair straight or wavy and not woolly. One even cross of a woolly-haired with a straight-haired race would hardly have transmitted such straightness of hair to posterity, but if after a first cross the fresh invasions, though only in scant filtrations, were of straight-haired people, the effect of the mingling of two straight-haired races with one whose hair was woolly would surely be to make the spirals uncurl.

Mr. Threlkeld, whose acquaintance with a New South Wales dialect seems to have been very thorough, denies† that the Australian language has any close affinity with the Malay either

* Grey's "North-west and Western Australia," vol. i. p. 254.
† "Key to the Structure of the Aboriginal Languages," p. 82.

in words or construction. This is a somewhat vague statement, in reply to which it can be said that, although the traces of Malay influence on Australian language are not numerous, they are unmistakable, and are observable at widely distant places. Duplication is met with in the Tasmanian, and may have been inherited in Australia from the primitive Papuan; it is not improbable, however, that its excessive use on the continent is due to the Malay, which reduplicates to form the plural. Often existing side by side with the form ' ngai,' a very prevalent alternative term for the first personal pronoun is ' adhu' or ' atoo,' which may perhaps be the equivalent of the Malay ' aku.' But I prefer to regard ' adhu' as an inflexion of ' ngai,' designating the subject when an agent. In the extreme north-west, where Malay words might be most naturally expected, very few are distinguishable. Mr. Curr has noted ' unbirregee' at Port Darwin and ' engeegee' in the Coburg Peninsula as the analogues of Malay ' gigi,' meaning *teeth*. It is rather at unexpected places that Malay words turn up, indicating either that the Malay inroad, if made at the north, took place in long past ages, or that now and again parties of Malays, either from choice or necessity, landed and became naturalised at various spots on the east, north, and west, and modified the speech of the people first immediately round them and then landwards. There are throughout Australia, in the main, two types of terms for *father*—a ' bapa' type and a ' mama' type. As there are similar words for *mother*, it might at first be conjectured that the terms for *father* and *mother* had become loosely transposed. I once thought so, but from the localities in which a particular type of term for *father* occurs, and from the occurrence of certain words in conjunction with the different typical ' father' terms, I have come to regard the ' bapa' type of terms as a mark of Malay descent, and the ' mama' type as equally evidential of great predominance of Papuan blood. Thus, for example, speaking generally, the dialects of Victoria * and West

* In the chapter on Language it will be seen that Rev. Dr. D. Macdonald (" The Asiatic Origin of the Oceanic Languages ") regards a word ' mama' as the vocative of ' abab' *father* in Efatese. It is scarcely credible that the corresponding Australian forms are thus related. They do not occur together. The Maar people who have wedged themselves in amongst

Australia, which are among the most pronouncedly Papuan,
are characterised by the 'mama' form; along the coast of New
South Wales and the eastern coast of Queensland, and for some
considerable distance inland, localities which, as I shall show,
possess unquestionable Malay words, the 'bapa' type of terms
prevails; whereas in Central Australia there is great variety of
terms interspersed with 'bapa' forms, but without a certain
recognisable third type, unless it be the 'nunchun,' which is
very probably primitive Papuan.   The Dravidian word would
approximate closely to the Malay, and it would, therefore, be
impossible to say with exactitude that a particular 'bapa' term
was Malay and not Dravidian, but the closeness of likeness
to the original Malay, and the·concurrence of other words
certainly Malay, will raise a presumption in favour of a Malay
lineage.

According to Marsden, the Malay 'mana' is properly the
adverb *where*, but is used idiomatically to signify *who, whom,
which, what*.   In many Australian words used interrogatively
'min' is a radical syllable.   It might, indeed, be said that
'mina' or 'minya' is an interrogative stem.   In the Kabi
(Queensland) dialect, for instance, we find 'minyanggai,' *what;*
'minyama,' *how many;* 'minyanggo,' *how;* 'minani,' *why*.   In
the Kamilroi, according to the Rev. W. Ridley, 'minya' signifies
what, and 'minyunggai,' *how many*.   At Barraba, 'menari' is
Kamilroi for *where;* at Port Macquarie 'minar' stood for both
what and *where*.   The Murra-worry tribe, between the Warrego
and Culgoa Rivers, employed the word 'minyan' to mean *what*,
and 'minyangor' to mean *why*.   Even to the north-east of Lake
Torrens, in South Australia, this class of interrogative is found.
This Australian word may be cognate with Semitic 'mi,'
'mah,' Heb. 'man,' Syr. 'ma,' Arab.   In no parts is the Malay
type of term for *father* so general and so stereotyped as
in conjunction with the etymon 'min' in interrogatives.   But
strangest of all is the occurrence of the word 'tungan' (spelt
also 'tongan' and 'tungun'), *hand*, which is evidently the
Malay 'tangan,' *hand*, also in the extreme east, and there alone
in Australia.   This most interesting fossil is found on the basins

the Kuli in the Western District, Victoria, preserve the 'bapa' type
of New South Wales, while the Kuli use the 'mama.' The difference
is one of the distinguishing marks of their respective dialects.·

of the Nerang Creek and the Tweed and Richmond Rivers, at the extreme easterly point of the continent, and reminds one of some great boulder carried by an iceberg from a high latitude thousands of miles from its parent rock and deposited when the iceberg has been overset or dissolved.

There is yet another not much less astonishing relic of Malay speech near the same quarter, and nowhere else so distinctly— viz., the word for *head*, which in Malay is 'kapala.' In New England the analogue is 'kopul,' on the Hunter River it is 'gaberong,' at Sydney it was 'kabura,' on the Castlereagh it is 'ballang,' on the Bogan 'bula'; and surely a better example of a contiguous group of terms, derived unconsciously from 'kapala,' could not possibly be given. The Malay word is the model of which the others are imperfect copies; it is the bull's-eye fired at, the others are the spots hit, some on one side of the centre, some on the other. The word for *skin* is also probably Malay, in which language the equivalent is 'kulit'; while in the east and south of New South Wales the term used is some such form as 'yulin,' 'ulan,' 'yoolak.' It might be assumed that two or three Malays were handed as human curiosities from tribe to tribe and found a last asylum near Point Danger, but the concurrence of five such indisputable glossarial vestiges suggests rather that there was a strong infusion of Malay blood added to the Kamilroi and allied tribes.

A track across the centre of Australia from the Gulf of Carpentaria southward is marked by a few Malay words of which the following are examples: 'kako' (Hamilton River) *elder sister*, 'kahkooja' (Darling River), *elder brother*, 'kaku' (Evelyn Creek), *elder brother*, corresponding to Malay 'kaku' *elder brother* or *elder sister*; 'Kutchiloo,' 'kichalko,' &c. (Darling River), *small*, cf., 'kechil,' 'kachil' (Malay), *small*. With 'kutta' (Daiyeri, S. A.), *louse*, compare 'kutu' (Malay), *louse*.

Another region where unquestionable Malay lingual traces exist is a tract on the east coast of Queensland, from about 17° to 21° S. lat., and inland to a distance of some two hundred miles. Three words diffused in this locality are distinctly of Malay origin—viz., those for *father*, *moon*, and *rain*. In Malay they are respectively ' bapa ' (Javanese ' baba '), ' bulan,' ' hujau ' (Javanese ' hudan '). The first is represented by forms such as ' baby,' ' babai,' ' abah,' ' yabba '; ' bulan ' has analogues in 'bullanoo,' 'balano,' 'pallanno,' 'palanoo,' 'bulbun,' and resembling the Malay word for *rain* ('hujan') are the following: 'Yookun,' 'hugun,' 'ukan,' 'yugan,' 'yukan.' The Australian words are certainly echoes of the Malay. In the same locality, with perhaps Halifax Bay as focus, I find two more words of Malay derivation occurring, and nearly as distinctly recognisable. The Malay for *bone* is 'tulang,' and for *house* 'rumah.' Equivalents about Halifax Bay and neighbourhood are 'toola,' *bone* and *wood* (Western River); 'toa,' 'tulkill,' 'tolkul,' mean *bone*, and all over that part of the country the word for *wood* takes such forms as 'tula,' 'toolani,' 'tular.' It should be noted that the Australian dialects frequently apply one and the same designation to *bone* and *wood*. The Australian words corresponding to 'rumah' occur at Halifax Bay, where 'ringo' and 'rongo' are used in the sense of *camp*. I would not adduce this as an analogy but for the preservation of the initial 'r,' a comparatively rare initial in Australia, and an anomaly in this particular spot.

On the Cloncurry River emptying into the Gulf of Carpentaria the word 'waramboo,'—spelt also 'ooramboo'—is probably a corruption of Malay 'rambut,' both meaning *hair*. At the same place, and only there, the term for *sun* is 'muntharra,' which comes very close to the Malay 'mata-ari,' and not very far distant—at Burke Town—the Malay 'bulan,' *moon*, has been the parent of 'ballanichi,' the word now in use for *moon*. The general term in West Australia for ear is 'twink' or 'dwonk,' which is most probably the Australian form of Malay 'duwan,' also meaning *ear*.

Besides these outstanding examples of Malay influence on the language, occurring at places so far separated, others might be instanced, the origin of which is less clear but probably Malay, and no doubt future research will disentangle many more words similarly derived.

There is proof positive that the best cave paintings have been executed by people of Malay blood from the island of Sumatra, a strong presumption also that the rite of circumcision was derived from the same people and place, and I am disposed to think that the Australian message-stick is a childish imitation of Malay writing upon bamboo and rattan cane as practised in Sumatra. These views will be stated at length and supported in the proper place, and if they be well founded the extensive prevalence of the practices referred to attests how powerful the Malay influence has been.

Before proceeding to a new department of inquiry, it will be well to recapitulate the view of the origin of the Australian race taken by the writer. Australia is first contemplated as occupied by a Papuan people, probably both sparsely and unevenly distributed. It is not affirmed that they were purely Papuan; the Negrito and the Melanesian may both have been represented and fused together, but for want of sufficient evidence this point is undetermined. Whence they set out and the route by which they came to Australia cannot be discovered; but, taking it for granted that the cradle of the human race was in Asia, whence all the nations have radiated like successive swarms from a parent hive, then the indigenes of Australia would most naturally come from the north and by way of New Guinea. The lineal descendants of the original Australian natives were the now extinct Tasmanians, who crossed from Victoria perhaps on dry land. Their migration from Victoria is held to be fairly established by the manifold forms of evidence already cited.

Australia is next regarded as invaded by a more advanced, straight-haired race which arrived at a very early period of the world's history, entered by Cape York Peninsula, and poured into central Australia with a general south-westerly current. Partly driving forward, partly cleaving, partly darkening itself by the tide of life upon which it presses, this stream inundates the whole country, but not to an equal depth.

Finally, another invasion takes place, also from the north, first with some degree of continuity, and then intermittently. This straggling stream winds about here and there, touches the shore at various places, and recoils back inwards. Indeed, this last influx may have come by several little rills, entering at places widely apart, and gradually losing themselves in the life-lake, as Austral rivers, exhausted by percolation and evaporation, disappear in the central plains. Australia is thus like a great lake which has been first filled by water of a particular tinge, and into which a clearer stream flows, crossing the lake, remaining purest in the course of its main current, then eddying hither and thither, and leaving the original water least altered in the bays around the margin. After receiving additions of water of yet another hue from numerous little surface rills at different points, the places of ingress are closed, the water stagnates, and the problem is to distinguish the different constituents in the lake's contents, assigning to each its place and relative proportion.

Upon the Papuan aborigines the Dravidian influx made a deep and general impression; the influence of the final arrivals, the Malays, was slighter and more partial. The first tenure by the Papuans, and their subsequent dispersion and dispossession, qualified by partial absorption, are shown by the relation of the Victorians to the Tasmanians, and also by the fact that a more particularly Papuan people fringes the coast, especially on the north, south-east and west. For example, there is an element in the Victorian tongue which is much more akin to the language of the people of the extreme west of Australia than to intermediate dialects. The following words are illustrative of this agreement :—

English.	Victorian.	Western Australian.
Father . .	maam, mama	marm, mam, mama.
One . . .	kaiap	kain.
Ear . .	wirn, wing, wiring	weening.
Sun . . .	nowingi, ngwingi, ngawi	nanga, nganga, nonga.
Wife . .	layarook, leurook	yokka, yawk, york.
Walk . .	yanna, yan, yungan (this word is common in east, south, and west of Australia, but not in centre)	yenna, yangwa.
Opossum . .	wolangi, wilang, wille, wollert	wallambine, wolumberree (occurs towards the north).

Besides agreement in particular vocables, there is a strong likeness in phonology. Then further, the word 'lar,' signifying *tooth*, is found in Victoria; and we have to cross the continent to the Gulf of Carpentaria before the same type with initial 'l' presents itself again, which it does at Caledon Bay in the word 'lerra.' Several words in 'l' may be taken as a class which serve to link together people in the extreme north who have been disparted by a wedge of linguistic influence forced in between. By this, as well as by Papuan physical characters being more pronounced on the coast, is the Papuan coast-fringe attested. The peopling of Australia, in so far as the succession and distribution or commingling of different races is concerned, has been not very unlike the settlement of Great Britain. The Keltic element in Britain is represented by the Papuan in Australia, the Saxon by the Dravidian, the Norman by the Malay. In both cases population has poured in mainly on one side, the earliest settlers gradually retiring to the farther shore. The second race takes entire possession of the centre, shedding the indigenes to either side. Wales and Cornwall might correspond to Victoria, the Isle of Man to Tasmania, not in relative position to the mainland, but in isolation and racial purity; and the Highlands of Scotland would represent Western Australia. In each case from the first two races the bulk of the people is sprung and the vocabulary and grammar are inherited, while the third race, sprinkled here and there over the land, has left the slightest lingual traces of its presence.

CHAPTER V

DISTRIBUTION

HAVING outlined the relationship between the different races who have settled in Australia, and indicated approximately where they first reached the continent, I shall now offer some observations upon the distribution of the population. Mr. E. J. Eyre propounded the theory that the aborigines reached Australia on the north-west coast, and settled it by spreading in three main streams—one by the west coast, another by the north and east coasts, a third crossing the centre southwards—and all three meeting again at the southern coast. This theory is adopted by the writer of the article on Australia in the "Encyclopædia Britannica"; it is also accepted, elaborated and strongly confirmed in Mr. E. M. Curr's work. I once entertained this view, but have been compelled to abandon it. I accept the evidence but reject the conclusion. A theory so deep-rooted and widely current, plausible and yet erroneous, demands strict examination, and if false, careful refutation. That the progress of settlement was from the north southward, and not *vice versâ*, is incontestable. That there are strongly marked differences distinguishing what Mr. Curr calls the Eastern, Western and Central Divisions may also be admitted.

Yet these premises do not lead to Eyre's theory of settlement. Mr. Curr's reasoning is vitiated at the start by his unwarrantable assumption that the Australian race is homogeneous. Had he believed that the Tasmanians were the first occupants of the continent, he would have had a powerful factor to account for differentiations, which cannot well be accounted for when the existence of an autochthonous basal race is ignored.

The outstanding characteristics of the three assumed divisions or streams of population have been indicated as the existence of circumcision and what may be called *concision* in the Central Division, the absence of these and the practice of naming tribes by negatives in the Eastern Division, and the utter absence of these three peculiarities in the Western Division. A mere *primâ facie* view of these distinctive features gives a bias against the three-stream theory, which does not even pretend to account for the rejection of the practices named where they are rejected; and, in fact, it is inconceivable that so strongly marked practices would have been abandoned in the districts where they do not obtain had they at one period been characters of the original stock.

The Tasmanians knew nothing of circumcision or concision; neither did the first-comers of the second immigrating race. Circumcision has been introduced in the north by the influence of Malay Mahometans in comparatively modern times. Concision,* or the "terrible rite," as Mr. Sturt called it, probably came after, and was gradually developed for personal adornment. These rites spread rapidly southward, and would, no doubt, have overrun the whole continent but for the advent of Europeans. Hence the prevalence of these rites tells us nothing about whence the aborigines came, nor how they were distributed. How or where the naming of tribes by negatives was introduced is an

---

* This name was suggested to me by the excessive extent of mutilation. Mr. W. E. Roth, in his most admirable work, "Ethnological Studies among the North-West-Central Queensland Aborigines," pp. 177 *et seq.*, gives a full description of the operation which he calls *introcision*, a name equally applicable to a corresponding mutilation of females. Prof. W. B. Spencer and Mr. F. J. Gillen call the rite *subincision*. They give a legend of the Arunta tribe, Central Australia, to the effect that shortly after men of the 'little hawk' totem had introduced circumcision by means of a stone knife, 'individuals belonging to the Achilpa or "wild cat" totem introduced the rite of Ariltha or *subincision*." Proceedings of the Royal Society of Victoria 1897. p. 146.

enigma.   It may have characterised the early Papuans; it may
have arisen through there having been a greater multiplicity
and confusion of tongues in the east than in the centre and
west of Australia.   Its occurrence in the east favours as much
my theory of settlement as it does the other.   Elsewhere I
show that this form of nomenclature probably originated in the
frequent utterance of negatives or corresponding words expressing ignorance of what was addressed in a foreign dialect.   The
linguistic evidence which Mr. Curr offers does no more than
support the hypothesis of a general movement from north to
south in the central part of the continent, which I also
affirm.

Having shown the invalidity of the inferences in favour of
settlement from the north-west by three streams, I shall now
adduce the proof that the migration was from the north-east
south-eastward on the east coast, southward, south-westward,
and westward elsewhere.   The linguistic evidence for this hypothesis is, I think, irrefragable.   It may be summed up thus: A
number of important words in the south-east, south, and west
of Australia may be traced through numerous modifications,
and the traced lines are found to converge about the base of
Cape York Peninsula, in the north of Queensland.   In fact,
some of them can be run right across to New Guinea.   In no
other locality can the languages be thus run to ground, as it
were, so that the *Igdrasil* of Australia may be said to have its
roots in the Cape York Peninsula.   Besides individual words,
other linguistic features can be traced to the same locality.

One of the most instructive lines is that which may be
formed by various names for the *emu*.   At Albany, in the extreme south-west of Australia, the term is ' waitch,' and in the
immediate neighbourhood the ordinary form is ' wadgie '; on
the Great Australian Bight the term is ' warritcha.'   On the
eastern watershed of Lake Eyre such forms as the following are
found: ' waraguita,' ' warrawatty,' ' wargutchie '; in the South
Gregory District, Queensland, ' warukatchi ' is found ; ' woitté '
is the term for *big* at the Coen River, flowing into the Gulf of
Carpentaria, near Cape York.   On Prince of Wales Island, in
Torres Strait, ' ure ' means *bird;* in New Guinea the term is
' ori '; and on Saibai Island, on the New Guinea coast, immediately opposite Cape York, ' uroi ' means *bird,* and ' kasa ' or

' kaiza,' *large*. To derive the West Australian word ' waitch ' we need to traverse the whole extent of Australia from the extreme south-west to its most north-easterly point, and then cross Torres Strait to the New Guinea coast, where we find its etymology in two words meaning *bird, large*. It should be observed that a pure sibilant is so rare in Australian languages as to justify the doubt whether it is used at all. Hence an ' s ' or ' z ' in New Guinea would become a ' t ' or ' ty ' or palatal ' ch ' in passing to Australia.*

It will be necessary for me to direct attention here to a linguistic phenomenon in a number of Australian dialects which is somewhat puzzling. I refer to the frequent change of initial ' w ' or ' wh ' into ' k ' and occasionally ' ku,' or, as it sometimes seems, the prefixing of a ' k ' before either ' o ' or ' u,' which would otherwise be initial vowels. In some cases this peculiarity might be sufficiently accounted for by supposing a natural relation between ' k ' and ' w,' whereby the latter may insensibly merge into the former or *vice versâ*; *cf. war* and ' guerre,' *ward* and *guard*; but in a number of cases it seems pretty certain that the initial ' k ' or ' ku,' and even perhaps ' kura ' or ' kuru,' as in the numerals, possesses or did possess a definitive value.

The derivation of ' waitch ' explains many of the names for *emu* scattered throughout Australia. Thus the word ' korre ' at Adelaide, South Australia, is probably just *the bird*; and at Kulkyne, on the Murray, Victoria, ' karawingi ' is the local equivalent for ' ori kaiza '; near Ballarat, Victoria, it occurs in the local name ' koraweinguboora.' A common Victorian form is ' kowir.' In New South Wales and part of Queensland it has been corrupted into ' ngooroon ' and similar forms.

There are practically only three names for *emu* in the west of the continent. The first is ' waitch,' already dealt with ; the second, ' yalliberri,' is found from the Murchison River northward to the Shaw River. Evidence is wanting for tracing this across the continent, but it prevails widely in the north-east, and the lines of prevalence are focused about the western watershed of the Burdekin in such forms as ' koolpurri,'

* "The sounds of *s* and *z* are wanting in Gudang (Cape York dialect), and when occurring in a foreign language are represented by *ch* or *ty*" (McGillivray's " Voyage of the *Rattlesnake*," vol. ii. p. 282).

'goberri,' &c.   'Kool-' is just a variant of 'kuro,' *bird*; and
'-purri' is an adjective signifying *many* or *large*.   The third
West Australian form is 'kullia,' occurring on the outside of
the territory where the other words prevail.   This is the local
form of the Darling 'kulti'; 'kul' corresponds to 'kool' and
'war,' and the termination suggests a decayed adjective.

The above derivations supply the key to the derivation of a
number of other bird names.   Many of the names for *swan*,
eaglehawk, and *native turkey* have the same meaning as the name
for *emu*.   The 'waldja' and 'warlik,' *eaglehawk* of West Australia, are both derived from 'ooreytella,' 'korytella,' 'koretalla,'
and similar variants of the north-east, the original form being
'koritalka' or 'oritalkai'; 'talkai' is a common term in the
north-east for *big*.   This inference is borne out by the name
for *emu* in Gudang (Cape York) being 'nichulka.'   'Talkai' is
probably contracted from 'talkari.'

As further illustration of the mode of forming compound
words and the accuracy of the above derivations, I may point
out that in one place 'oorumpa' is the name for *wild turkey*, in
another 'oorooba' is the name for *emu*; in both instances 'oor'
corresponds to 'ori,' *bird*, and 'umpa' is an adjective in various
dialects meaning *big*.

The equivalents for the numeral *one* occurring throughout
the greater part of Australia can also be traced to the north-
east, and three distinct Australian forms are discoverable in
New Guinea and adjacent islands.   In West Australia the
prevailing vocable for *one* is 'kain'; in Victoria it is 'kaiup' or
'kaiap.'   The affinity of the two is suggested by their resemblance.   The one is corrupted from such a form as 'koornoo,'
occurring in the north-east; and the other is the local variant
of a very widespread form, the type of which is 'kurupa,' with
sometimes a final 'na.'   An example of the fuller form is
'koorbno,' on the Diamantina River, Queensland, which, when
compared with other northern and New Guinea forms, leads to
the above conclusion regarding the etymon.   Victorian 'kaiup'
is traced northward thus: Bumbang, Victoria, 'geyabi'; Wal-
jeers, New South Wales, 'kooinebine'; Wellington, New South
Wales, 'oonboyie'; Castlereagh River, New South Wales,
'ngunbeer'; Diamantina River, Queensland, 'koorbno.'   West
Australian 'kain' has resulted from such changes as follows:

Great Australian Bight, 'kean' and 'kyunoo'; Lake Eyre, South Australia, 'koono' and 'koornoo'; Cooper's Creek, 'koornoo'; Diamantina River, Queensland, 'koorbno.'

The term 'kuma' or 'kooma,' *one*, of Adelaide and neighbourhood, can be traced northward in the same way. At Mount Remarkable it is 'kooman'; Gawler Range, 'goo-o-mana'; eastern shore of Lake Torrens, 'koopmana.' We then reach the region where 'koornoo' has held its place, but at the head of the Hamilton River the form is 'gooniba.' The 'm' in 'koopmana' has probably crept in from its relation to 'p.' I think there can be no reasonable doubt that both 'gooniba' and 'koopmana' are variants of Diamantina River 'koorbno.' Another set of forms is traceable from Melbourne northwards along the coast to Gladstone, in Queensland, thus: Melbourne, 'karnboo'; Gippsland, 'kutupona'; Stradbroke Island, Queensland, 'kurraboo'; Burnett River, Queensland, 'karboon'; Gladstone, 'karboon.' I do not affirm that this last has necessarily been conveyed southward continuously along the coast. The different forms may have reached the coast at the various points, in the speech of natives that had parted in the north, but obviously they are all derived from the inferred original type 'kurupona' or 'kurupana." On the north-east coast of Queensland the most prevalent term for *one* is ' woorba,' which can be traced northwards to Saibai, on the coast of New Guinea, in the following series: Peak Downs, 'woorba'; Rockhampton, 'werpa'; Mackay, 'warpur'; Belyando River, 'wirburra'; Port Denison, 'warpa'; Prince of Wales Island, 'warapune'; Warrior Island, 'woorapoo'; Saibai Island (New Guinea coast), 'urapon'; Bula'a, New Guinea, 'koapuna.' Other forms in the north-east of Queensland are 'noobun,' 'nupun' and the like, represented in New Guinea by 'obuna,' 'abuna.' All the above belong to the one type. Another type, of which the etymon is 'kueitan,' can be traced from Victoria through New South Wales and Queensland, also to the north-east coast, and the corresponding form 'koitan' is picked up on Woodlark Island to the east of New Guinea. The stages of change may be briefly indicated thus: Piangil (Victoria), 'yaitna'; Tintinaligi (New South Wales), 'ngitya'; Cooper's Creek, 'waityu'; Paroo and Warrego, 'itcha'; Mackay, 'watchin'; Belyando, 'wogin'; Cape River, 'whychen'; Woodlark Island, 'koitan.' This treatment of the argument from the distribution of numerals will have to suffice for the present.

A very important mark for relating tribes and dialects is the term used for *man*. The sum total of these is not large, and with few exceptions they can be traced also to the north-east of the continent. Thus in 'kerna,' on the Hamilton River, in the north-west of Queensland, are focused 'koori,' Hawksbury, New South Wales; 'kouai,' Gippsland; 'kooli,' Victoria generally; 'korni,' mouth of Murray River; 'kurda,' Streaky Bay, South Australia; 'karoo,' Shark's Bay, West Australia; 'kurna, Cooper's Creek. The name 'maar,' or 'marra' (Warrambool, Victoria), is traceable through New South Wales and Queensland, and appears on and near the Queensland coast on the north-east as 'mari,' Port Denison; and 'murree,' Porter's Range. Evidence of this special kind might be multiplied, but I forbear. Further corroboration of the north-easterly origin of the natives is furnished by the fact that a peculiar form of dialect found in the very heart of Australia at Alice Springs and neighbourhood is most closely related, phonologically, by vocabulary, and by the exceptional feature of aversion to initial consonants with dialects at the Norman and Palmer Rivers, near the south-east corner of the Gulf of Carpentaria, and, singular to say, with the Gudang at Cape York.

Again, a particular type of pronoun prevails throughout almost the whole of Australia. It is more or less mutilated in New South Wales, Victoria, South Australia, and West Australia, and its most perfect types, so far as yet made known, are found in the Kabi and Turrubul of Southern Queensland, the Kaurarega of Torres Strait, and the Saibai, near the New Guinea coast.

A striking fact emerging upon philological research is that two currents of language have actually crossed each other in the east central part of the continent about the neighbourhood of Cooper's Creek. The Comparative Table in this work will show that a stream of population has crossed the continent from the Cloncurry River, a tributary of the Flinders flowing into the Gulf of Carpentaria, direct to Adelaide and neighbourhood in the south. The Darling River blacks are perhaps the main representatives of this migration. The field occupied by dialects of this east central type has been cloven by the language of one or more streams of people passing down the rivers from the north-east, the Diamantina, Thomson, Barcoo, Booloo. The centre seems to have been first occupied and the later streams seem to have forced their way westward and to have formed the almost homogeneous people of the extreme south-west before circumcision or "the terrible rite" was introduced. This strange phenomenon has not been noticed before. We do not possess evidence to trace the northern end of the east-central current back eastward farther than the Cloncurry River on the south of the Gulf of Carpentaria. I do not doubt that it will yet be traced northward into Cape York Peninsula. If it was prior to the westward current it would probably be pressed to the west as well as cloven by the new immigrations coming from the north-east coast; which stream first occupied the centre I cannot with certainty determine. Lake Eyre is a meeting-point of northern, western, and east central divisions.

In the east of Australia, the territory which on linguistic grounds I have divided latitudinally into two divisions, there is very clear proof that a double line of advance from north to south was made, the dividing-line corresponding roughly with the Dividing Range. There are thus two longitudinal sections, the coast one terminating in Gippsland, the inland one terminating in the rest of Victoria and the south-east corner of South Australia.

The problem of the intermixture of races may also be dealt with by determining to what race belong the extremes when viewed as regards the time of their arrival. The first-comers being Papuans and the latest arrivals being as distinctly Malays, an intermediate residuum which is neither the one nor the other, and which yet has contributed most largely to the population, requires to be accounted for.

The pronouns and numerals are the main distinguishing linguistic features of this racial element, which may be called the Australian proper. The peculiar type of pronoun in a more or less perfect form covers nearly all Australia. It is very distinct and well elaborated. It differs from the Tasmanian pronoun; it is certainly not Malay. It approaches closest to the Dravidian, and along with other marks justifies the inference that the predominant element in the native Australian race as now existing is constituted by descendants of a people allied to the aboriginal race of Central and Southern India.

CHARACTERISTIC AFFINITIES BETWEEN LANGUAGES OF WEST AUSTRALIA, SOUTH AUSTRALIA, QUEENSLAND, AND NEW GUINEA, AS PRESENTED IN A ZONE PASSING FROM S.W. TO N.E.

English	S.W. of West Australia	Coast of Great Australian Bight	Lake Amadeus	E. Watershed of Lakes Eyre and Torrens	Upper Basin of Paroo and Warrego	N.W. Central Queensland	Burdekin Watershed	Coast, or near it, N. of Burdekin	New Guinea
Dog	dooda, doodoota	—	—	toota	—	—	oodoodoo	oota, kaia, whoyyer	wasi, kwai waeha
Crow	wordong, wardung	—	—	worder, wardergun	woterkan	—	wethergun, waw-teringa	wutthagun	—
Emu	wadgie, waitch	warritcha	—	waraguita, warrawatty, warrawidgee, wargutchie	—	wakaje	—	korundi koorangea	ori, uroi, *bird.* kaiza, kasa, *big*
Eaglehawk	waldja, warlik	walya	wollowara	wirdla, kaldura, kurlathura	koothalla	wollayun, kooridala	wirta, koorathul-la, ooreytella	korytella	—
Egg	buoya	kabin, namboo	ambo	karboin, kapi, pampu	kobwee, kaboin	bambo	kookabinyu	bamboo	kap, pawu, pou
Fish	wappie	—	—	koppi	—	koopa, kope	—	wappi	wappi
One	kean, kain	kyunoo	—	koono, koorno, koorbno	—	—	—	nupun	urapon, koapuna abuna
Two	kootera, goochal	kootera	godarra	kootera	wootah	uttera, kotya	kotoo	—	okasara, uksara,
Three	mow, murdine	—	munkurippa	—	—	worita, mulia	—	murgine, mundula	—
Mother	ngungan	—	yackhoo	yunga, younginna	—	yunga, yungan	yungana, younga	—	—
Head	katta	—	cutta	karte	—	kata, kanta	katto	kutthul	—
Mouth	da, dow, dawar	—	tar	towa, tower	—	unta	—	thowa	—
Teeth	ngulla, nulga, ngorlok	—	—	nulgulla, nalkulla	—	milka	nulla	—	—
Beard	nganga	—	ungurra	nunka, nanga	—	ungka	nunga	nunga	—
Stomach	kabool	—	—	yaboo	—	—	knambooma	kippa	—
Blood	knooba	—	—	kooma	kooma	gimpa, kimba	kooma	kooma	—
Tomahawk	kodja	—	—	goocha	—	kooya, *knife*	koocha	manyi-gogeo	—
Rain or water	kaip, gabby	koppe	cobbie	kapie, knappo	koommoo	uppo, yuppo	kammoo	kumoo, kap-par	—
Wood	boona, bono	—	—	bunda	baka	—	boorri, budda	batchu	—
Stone	booya, booyee	boory	—	berry	burre	poori	burrey, barrie, byee	burry, byree	—
Dead	wannega	—	illoong	woolinga	woolul	—	woolinya, woo-nunga	oolunga	kwarea, wareha, walega, *die*

This theory is held in conjunction with the belief that on the north-west, north, and north-east coasts there have been desultory landings of small bodies of people not in the main currents. There are indications of groups of Melanesians having reached Australia on the eastern Queensland coast perhaps as castaways, and having penetrated inland, leaving their impress upon the practices and language.

A table is subjoined showing that a large proportion of West Australian words have crossed the continent by a south-westerly route from the north-east of Queensland.

When we keep in mind that a Papuo-Tasmanian influence survives specially in the south, on the west coast and on the north coast of Australia, which cannot possibly be traced to a point of first arrival on Australia, the linguistic evidence given above is so varied in character and so massive in quantity, and in several cases exhibits withal the gradual transformation of words so distinctly, that it leads irresistibly to the one conclusion, viz., that the chief of the three easily distinguishable elements in Australian language entered Australia on the north-east, and the inference is inevitable that the people who spoke that speech passed to Australia from New Guinea.

CHAPTER VI

PHYSICAL CHARACTERS OF THE AUSTRALIANS

Physical, mental, and moral characteristics—Physical appearance—Mental characters—Ramahyuck school *one hundred per cent. of marks* for three consecutive years—Imitation—Moral characters—Instability—Sympathetic and affectionate—Gaiety—Improvidence—Native police—Missionary effort—Barbarous whites, humane pioneers.

THE physical appearance of the natives is subject to considerable variation not only in different localities but even in the same community, and this as regards stature, muscular development, cast of features, and other particulars. Some of these differences are doubtless attributable to climatic influence, some to the difference of food products, while some are as certainly hereditary racial peculiarities. The wretched emaciated creature whose bones may all be told through his skin, although often presented to us as the picture of the Australian, is not a true picture. Such will be the appearance of parties where the food supply is always scant, or of others at a time of the year or in an unfavourable season when food is much more scarce than usual. It is also true that the inhabitants of the interior and the north are more spare, and perhaps on the average taller than those in the east, south, and west, but men of muscular frame and stout build are common enough in the coast districts other than the north. Taking the continent all over, the average height of the men will not exceed 5 ft. 6 in., and of the women 5 ft. There is, however, hardly a community in which two or three six-footers will not be found. As a rule, the muscles are not largely developed, but there are numerous exceptions. In Southern Queensland I have seen a type of man about 5 ft. 4 in. in height, thick set and powerfully muscular. One man of this stamp received his name from the massiveness of the calves of his legs. But even the lanker men are very strong and wiry in proportion to their weight, both bone and muscle being excessively tough.

The colour of the skin is shaded from a dusky copper to a brownish-black. The new-born babe is singularly fair, but becomes gradually darker with age. The natives have a predilection for ebony skins as a mark of beauty, a preference which may be due to the fact that the substratum of the population was originally darker. In those parts of the country which have already been particularised as more distinctly Papuan, there is usually an abundance of hair on the face and breast, a characteristic which accompanies increased squareness of build and greater muscularity. In the central parts there is less beard and less hair on the breast, and in the north, in some parts at least, the body is smooth and the beard very scanty. Throughout the continent the hair of the head, with some notable exceptions, is of a glossy raven black, very redundant and usually wavy. Where the Papuan blood is most predominant the hair is often curly and frizzy and sometimes woolly. I knew one black boy in the south of Queensland whose hair was of a dirty yellowish-brown, and there are several well-authenticated cases of true natives having hair that has been described, perhaps with poetic exaggeration, as golden yellow. A particular instance is given in the family of a man named Teacup, a leading blackfellow among his countrymen about Beemery Station, between Bourke and Brewarrina, in New South Wales. His children were copper-coloured and had long straw-coloured hair.* Such cases may arise from poverty in the black pigment, but seem too decided to be ascribable to such a cause.

There could hardly be a more striking contrast than that between the lank, tall, smooth, small-featured Northern Territory man and such a Victorian black as Bidhanin, well known at Ballarat under the name of King Billy. The latter was short of stature, not exceeding 5 ft. 4 in. in height, his hair hung in heavy wavy locks or tangles, his face was almost hidden with beard and whisker, and his bosom thickly covered all over with a dense crop of hair of two or three inches in length, so as to have quite a shaggy appearance. This man, born at Ercildoune, was a good specimen of, what I take to be, the Australian

* Informant, Mr. Colin Fraser.

Papuan. His features were also of the typical Australian
Papuan cast—*i.e.*, the brow comparatively low and retreating,
eyebrows prominent and shaggy, eyes fairly large, the iris being
dark brown and the white of a smoky-yellowish tinge, the nose
large and broad but not to say flat, indeed sometimes decidedly
Jewish, the nostrils wide, the mouth large, the lips thick, but

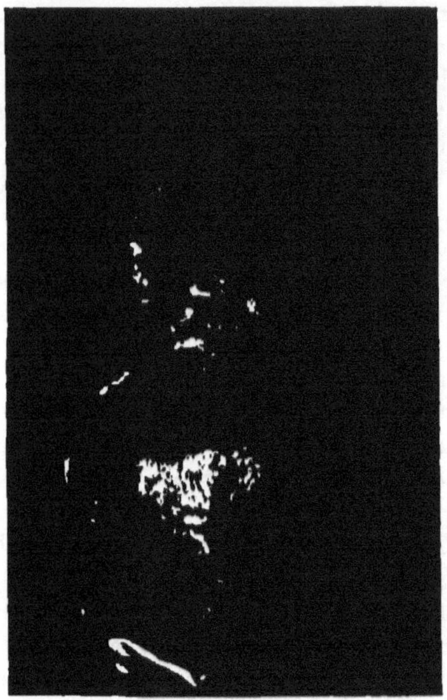

OLD PETER

without the swollen thickness of the negro lip, the cheek-bones
high, generally small and receding jaw, somewhat prognathous,
teeth large. This is the Australian Papuan face, and may be
met in many localities. I have a portrait of a black, known as
Old Peter, who belonged to Milroy Station, on the Culgoa River,
New South Wales. This portrait might pass for a presentment
of Bidhanin mentioned above. The trunk in front is completely
covered with dense hair, which spreads over the shoulders and
down the outside of the upper arm. The beard is thick, long,

and curly, with a tendency to fall in ringlets.   Old Peter was evidently stout and muscular for his height.

Alongside of people like those described there may be found others with features which might be called fairly good-looking, judged even by European standards.   These have quite a different style of forehead, narrow, smooth, rounded, high; also a much smaller nose, sometimes straight and full, sometimes snub and inclining to be tip-tilted, the lips full but not extra thick, and the facial outline a graceful oval.   A poor, unfortunate wretch of a black boy, who went by the name of Dougal, a native of Yabber Station, in the Wide Bay district, Queensland, would be one of the best examples of this latter type.   His face as a whole might have been called handsome.   He ended his days on the gallows-tree for crimes committed after he had become demoralised through the evil influences that blotch the gold-diggings.   Although the eyes of the Australians are rarely, if ever, oblique, a face with a decided Mongolian cast about the brow, cheek-bones and nose is occasionally met with.

There are certain peculiarities about the average Australian head which serve to mark it very distinctly.   It is of a pyramidal shape, the skull is abnormally thick, the cerebral capacity is about the smallest of all races.   Viewed in profile, the tip of the nose is the apex of an angle, the sides of which recede with about equal obliquity from a horizontal passing through that point.   The head is well poised, commonly having a backward lean, and is supported on a neck short and comparatively thick.

In general appearance the average Australian is symmetrically proportioned.   More bone and muscle would undoubtedly be an improvement, for a too common attenuation of limb and fineness of ankles and wrists are suggestive of weakness.   His hands are small and bony, the feet by no means large, seeing that they are always bare and used so much and in such varied ways.   The aboriginal is very strong for his weight, exceedingly agile, and has an erect, free, and graceful carriage.   As he is so largely dependent upon the exercise of his senses they are singularly acute.   His powers of tracking are proverbial.   My belief is that they are due as much to exercise as to peculiar natural capacity.   While in his native bush, all the blackman's senses are incessantly on the alert, it is therefore no wonder that his faculties of sense-perception should be highly developed.

MENTAL AND MORAL CHARACTERISTICS.

For a people so low in the scale of civilisation the Australians exhibit powers of mind anything but despicable. They are very keen observers, of quick understanding, intelligent, frequently cunning, but, as might be expected, neither close, nor deep, nor independent thinkers. In schools, it has often been observed that aboriginal children learn quite as easily and rapidly as children of European parents. In fact, the aboriginal school at Ramahyuck, in Victoria, stood for three consecutive years the highest of all the state schools of the colony in examination results, obtaining *one hundred per cent. of marks*. While among Europeans the range of mental development seems almost unbounded, with the blacks its limit is soon attained. An inherent aversion to application is generally an impassable barrier to the progress of an aboriginal's education; in addition to which there is usually the absence of sufficient inducement to severe mental exertion. Unless in the case of those who are so situated that they cannot help attending school, most natives who have been taken in hand to be taught have at best learned to read words of one or two syllables and to write their own names in a very clumsy manner.

A common feature in the aboriginal mental make-up is a propensity for mimicry. They are fond of imitating one another with a view to exciting ridicule, and they instantly seize upon salient peculiarities of white men, especially of strangers, and reproduce them with considerable success. It is astonishing how easily and completely young blacks, not cut off from intercourse with their relatives, but living and working constantly among the whites, fall into European modes of thought. To the influences of the white men they move among their mind seems to be a *tabula rasa*. Give such an aboriginal a white man's features and complexion and he is, to all intents and purposes, a white man of the unreflecting, uneducated class; some of them, with little or no incentive save the approbation of Europeans, falling into the routine work of the station, doing it with fidelity and pride, and for perhaps only a tithe of the white workman's reward.

In the aboriginal character there are many admirable, meritorious elements, but there is a lack of a strong, inherited, combining, marshalling will or self-determination, and, as a natural consequence, the moral qualities are prone to operate capriciously. The natives are not insensible to promptings of honourable feeling, but generally, unless when repressed or constrained by fear, they act from impulse rather than from principle, and their best inclinations are easily overpowered by pressure from within or from without. You could rely upon a blackfellow being faithful to a trust only on condition that he were exempt from strong temptation. One of the most condemnatory testimonies that ever has been given of this people is that which was given by Mr. Jas. Davies (known as Darumboi by the aborigines) before a commission of the Legislative Assembly of Queensland in 1861. As far as personal experience went this witness was well qualified to speak, for he had lived continuously among the blacks for fifteen years and three months. He said, very forcibly, " Hundreds of them would take your life for a blanket or a hundredweight of flour. I wouldn't trust them as far as I could throw a bullock by the tail." " They are so greedy that nothing can come up to them." " They are the most deceitful people I ever came across." " The father will beat the son and the son the father. The brother will lie in ambush to be avenged on the brother; if he cannot manage him in fight he will lie in ambush with a spear or a club."

This, I am sure, was stating the case against the poor creatures too strongly. They are not wantonly untruthful; they are not deficient in courage; they are not excessively selfish; and they are by no means lacking in natural affection. But Mr. Davies corroborates what I have said of the presence of that defect of character which may be termed instability. It may be said that the whole fabric of their moral character is in a position of unstable equilibrium. The slightest strain will destroy the poise.

They have a courage which fits them to perform marvellous feats of tree-climbing, gives them spirit to assert their rights in the face of danger from the white man's superior knowledge and strength, and, for a time at least, qualifies them to excel as roughriders. But their bravery is neither steady nor deep-rooted. No doubt they are very covetous, but they are also very generous. One of the nuisances which used to vex squatters was the good-natured recklessness with which a black boy would scatter about among his friends the rations or clothes he had earned by his own labour and which he needed for himself.

As a rule, the blacks are sympathetic and affectionate, especially the women. Sufficient evidence of this is the way in which white men have been treated who have been unfortunate enough to be cast upon their mercy. Relatives are usually fondly attached to each other. The attachment between parents and their offspring is very strong, and exhibits itself in kindness to the aged, who are tenderly cared for, and indulgence to little children. One case of filial affection which came under my own notice I cannot forbear to mention. It was that of a boy who had travelled with a stockowner to a considerable distance from his native place, showing his love to his parents, in a way very substantial for a black, by sending them a pound note through the post.

An almost universal feature in the aboriginal character is gaiety of heart. This, I believe, is a Papuan inheritance. Open light-heartedness was one of the pronounced features which Wallace observed distinguishing the Papuans from the Malays. Of the Papuan he says:* "They are energetic, demonstrative, joyous and laughter-loving, and in all these particulars they differ widely from the Malay." The open, sunny-hearted qualities are indisputably Australian characteristics. The Australian is good-humoured, enjoys a joke, and does not long harbour resentment. The absence of constraint in the direction of joyousness is accompanied by liability to unrestrained bursts of passion, which lead sometimes to most violent assaults. The black is a very vain man, conceited of himself and conceited of his countrymen, for reasons no doubt sufficient to him if not to us. It is perhaps as much owing to his vanity or his fondness for praise as to any other motive that he has been got to work at all. For, what other inducement is there to him to toil for the white man, and why should it not be to him rather a merit than a disgrace that, from our point of view, he is indolent? Has not nature dealt bountifully with him? If he makes his demands upon her at intervals with sufficient urgency, he may loll on her soft warm bosom at

* "The Malay Archipelago," p. 592.

his ease without discredit, until hunger compels him to stir. At light kinds of labour he can work well, and if it suits his purpose, he can apply himself diligently for a while, but, as he only has to provide for to-day, he does not trouble about to-morrow. He is not invariably and in every respect improvident, however. If he does not require to rob a bee's nest to satisfy present wants, he will indicate his discovery and assert his ownership by marking the tree which the nest is on, and will take the honey at some future time. In the Bunya Mountains in Queensland it was a common practice, when the Bunyas were in season, to fill netted bags with them, and bury a store in the gravel of a creek-bed, to be exhumed when required. The blacks of Western Australia store zamia nuts by burying in the ground, but without nets.* In these and various other ways the blacks show that they do not live an out-and-out, hand-to-mouth life. They are not cultivators of the soil, they neither sow nor plant (although I have known a black to plant a tea-tree in a locality where none was growing), but they reap grain, and roots, and fruits, preparing them in various ways for consumption.

Settlement by the British has usually proceeded without much resistance. The blacks have kindly assisted in their own dispossession and extermination, guiding the aliens through their forests, giving them much of their own strength at a beggarly rate of recompense, submitting contentedly to indignity and oppression, and rewarding injuries and insults with gentleness and service. They have committed robbery, rape, murder, and perpetrated several massacres. True, but they have often been trained to such offences by the lawless, brutal, indecent, tyrannical behaviour of the white men with whom they have come into contact, for as a matter of fact the outskirts of civilisation have a strong admixture of barbarism.

The first time that I saw a large number of blacks was at Durundur Station, sixty miles from Brisbane, in the year 1865. A bullock had just been slaughtered by the station hands, the blacks were congregated round the killing-place. A low white, with a feeling of gay superiority, swung the reeking, bleeding lights and liver with a slap round the neck and on to the naked bosom and shoulders of an unoffending black woman. The

* Grey's " Journal of Expeditions of Discovery," vol. II. p. 64.

F

gentle creature received this act of gallantry with a smile. I can never forget this disgusting insult and the meekness with which it was borne. It was at once an index and a type of much of the treatment which the natives have received from those who have taken their heritage away from them, and if the weaker side has retaliated is it to be wondered at? The cruelties perpetrated by the native police upon their own kindred in the name of law, although excessive and often unwarrantable, may be passed over here, because, granted the right to colonise and dispossess, a certain degree of conflict was inevitable, and it has been alleged by humane and competent judges, that where the native police, well-officered, patrolled a district, not only was property secure, but the blacks were exempted from vengeful and bloody attacks by the settlers. But woe for the lustful and atrocious conduct of individual white men, who, feeling secure from legal penalties and native reprisals, outraged and oppressed and hunted at their will. The small success of missionary effort, with which the unsettled life of the aborigines has had much to do, has led many people to conclude that they are not amenable to spiritual influence, and some settlers have adopted the fantastic, convenient, and self-exculpatory theory that the blacks have no souls. But, on the mission stations especially, there have been numerous proofs that the gospel appeals as much to an aboriginal Australian heart as to that of any other nationality, and that, notwithstanding instability of character, Christ is the power of God to the Australian.

It used to be a common maxim among bushmen, "It's no use to hit a blackfellow with your fist, he won't feel it," and the corollary was that a heavy boot, or a stout stick, or an iron bolt, or a stock-whip, were legitimate and suitable instruments for hortatory and punitive purposes. A powerful, heavy bullock-driver would maul a black boy as an elephant might a baboon; to kick the offender, trample on him, and kneel or tumble on his chest and stomach, were usual courses of procedure, and the brute who could do these things deftly and inspire a wholesome awe in the outraged would be entitled to respect. "I would as soon shoot a blackfellow as a dog," was no uncommon saying which some carried into practice. Concubinage was general, terrorising and murder, both by poison and bullet, plentiful enough on back stations, and used to be spoken about freely where not practised.   At the bar of God the souls of the aborigines will have a heavy indictment to present against men of our blood who have wronged and brutalised them.

While acknowledging and deploring the excesses of which the colonists have been guilty, it would be unjust to overlook the manifold instances of habitual humane treatment at the hands of some of the station owners and their employés.   But nothing deserving the name of an equivalent has ever been rendered, whether by individual favour or associated effort in civilising and Christianising, to the weak, peaceful, kindly people from whom Australia's glorious golden land has been wrested so speedily and at so trifling a cost.

CHAPTER VII

DWELLINGS, CLOTHING, IMPLEMENTS, FOOD

Dwellings, clothing, food, &c.—The blackfellow's home—His clothing
—Preparation of rugs—Use of bark of native tea-tree—Ornaments—
Cicatrices—Piercing septum of nose—Bags and baskets—Weapons—
Food, from cicada to kangaroo—Method of eating honey—Nardu and
nardoo—Bunya—Pitcheri, *combungie* or *wangle*—Ovens—Diseases—
Caused by sorcery—Treatment—Longevity.

THE home of the blackfellow is identical with the tract of
country over which he ranges; his dwelling is a structure of the
airiest, flimsiest kind. A breakwind of a few boughs proves
sufficient in fine weather, and in cold or wet he procures two or
three sheets of bark, sets them on end upon a crescent base-
line, one sheet overlapping another, the lap increasing upwards
so as to gather the sheets at the top. The whole leans upon a
few light props placed in front, the lower ends of which are
stuck in the ground, the upper ends converging and held
together by a natural fork in the end of one of the poles. This
description applies to the most common dwelling; sometimes a
booth of boughs suffices, while on the other hand rude little
cabins thatched with grass and mud are met with occasionally;
and near the most easterly point of Australia, probably owing
to Malay influence, the walls of the houses were of stakes inter-
laced with vines. The size of the house is determined by the
number of occupants it will have to accommodate in sleeping
posture. The floor is the green turf. The open front serves
equally for door and window. As the fire is lit far enough out
to allow plenty of room for the sleepers to stretch themselves
with their feet towards it, a chimney is unnecessary. In rainy
weather a small gutter is dug around the dwelling. Light huts
of this description are peculiarly suitable to a nomadic people,
unacquainted with metals, possessing few tools, and rarely ex-
posed to severity of climate. They may erect them in the first

instance with very little trouble, tenant them for two or three months at most, and then either carelessly leave them standing or lay the bark down flat and place a log or two on top of it to keep it from getting warped or lifted by the wind. If the bark is thus conserved, when the people revisit the locality the house is rebuilt in a couple of minutes. Such a structure constitutes a by no means uncomfortable sleeping apartment, and a residence commodious enough for people who can carry all their chattels with them; it has also this advantage, that it can be shifted as the wind veers and the open front be always on the lee side. When the natives were numerous their camps would contain twenty or thirty huts, and on the occasion of special gatherings there would, of course, be more. Each family would have its own dwelling. Young single men would sleep in groups apart from the families, and it is said that in some tribes the positions taken up by individuals were determined by considerations of kinship.

Almost the only real article of clothing worn by the Australians is the opossum rug. In the extreme north it is not in use. About the neighbourhood of Port Mackay, in S. lat. 21°, it is used,[*] but in Central Australia, right across the continent, the blacks are destitute of clothing. While travelling in the north-west Captain George Grey[†] saw no opossum rugs in use north of 29° S. The opossum rug serves equally well for mantle and blanket, and forms a receptacle on the mother's back in which she can carry her infant when on the march.

In making the rugs, the flesh is cleaned thoroughly off the skins, which are made pliable by rubbing with pieces of free-stone. They are generally ornamented with rude scratches representing snakes, emu's feet, and the like, the figures being coloured with red ochre. The skins are neatly sewn together, kangaroo sinews serving as thread. I was told by a black boy that his people in the Wide Bay and Burnett Districts, Queensland, were wont formerly to make the soft papery bark of the native tea-tree supply the place of blankets. It appears that the same practice obtains in the neighbourhood of Halifax Bay.[‡] At the Daly River, in the north-west of Australia, the same

---

[*] Curr's "The Australian Race," vol. III. p. 45.
[†] He is quoted by Mr. Curr, but I cannot verify this reference.
[‡] Curr's "The Australian Race," vol. III. p. 426.

kind of bark was used for many purposes. In many parts the females, and more especially young girls, wear a fringe suspended from a belt round the waist, the fringe being made of various materials, such as vegetable fibre, fur-cord, skin, &c. They are not excessively given to adornment of the person, but a few simple ornaments are very generally worn. Among these may be mentioned chaplets round the head, usually painted with pipeclay or ochre, strings of bright yellow reed beads, dogs' teeth and a piece of shell like mother-of-pearl suspended on a string worn round the neck. On certain occasions feathers are worn in the hair.

Ornamentation of quite a different kind is effected by raised cicatrices arranged in rows in various parts of the body. These are commonly made on the back, breast, abdomen, shoulders, and upper part of the arm, and only on males. The incisions are horizontal on the trunk and longitudinal on the arms. When first cut they are filled with ashes, charcoal, or some other innocuous material to keep the sides of the wound from closing, and to make them rise when healed, like a pair of lips.

In most tribes the males pierce the septum of the nose. All natives frequently anoint themselves with grease and charcoal. In fact, this anointing is practised on new-born babes, and is doubtless far more beneficial for infants than washing would be in their rude mode of life. On special occasions, such as man-makings, corroborees, and fights, the men smear their bodies with red and white clay in fantastic designs.

The women make bags of network, the size of the mesh as also of the whole bag being regulated by the use for which the article is intended. The cord employed in the manufacture is usually made of fur. Baskets, known by the whites as " dillie-bags," are woven of strips of cabbage-tree, tough grass, or the bast-bark of trees like the currajong ; a piece of cord is attached to opposite sides of the edge by its two ends, so as to allow the bag to be carried in the hand or slung upon the shoulder. These were the depositories of their valuables.

As regards weapons, I shall content myself with giving little more than a bare enumeration ; for a full and accurate description Mr. Brough Smyth's " Aborigines of Victoria" may be consulted. The characteristic and distinctive Australian weapon is, of course, the boomerang, which is made of various sizes and weights and shapes.   As already stated, similar weapons are used in Africa and India, but that which distinguishes one kind of Australian boomerang from every other is the property of returning to the thrower.*   In the south of Queensland the blacks had a very singular arm made of wood, about as flat as a boomerang, but considerably larger and heavier, and bent naturally at a right angle about the middle in the plane of its width.   It was probably an arm for close fighting, a kind of battle-axe in fact, although in outline proportioned more like a single-headed pick.   It resembles the leonile of Victoria figured in Mr. Brough Smyth's work, which was used in single combat.   Wooden spears are universal.   They are of diverse lengths and differ much in the design of the point, from simple sharpness to many barbs, sometimes cut out of the solid, sometimes of bone or flint affixed.   Some tribes make reed spears as well.   In many parts the spear is launched by the aid of a "throwing-stick" about two feet in length, now widely known by its aboriginal name *womera*.   One end of the "throwing-stick" is barbed, the tip of the barb rests on a hollow in the end of the spear and the other end of the "throwing-stick" is held in the hand.   This auxiliary, like the cord of a sling, increases the velocity with which the weapon flies.   The women in some communities have a special kind of spear about four feet long, called by the whites a "yam-stick," which they employ either for digging or for feminine duels, in which they are handled single-stick fashion, while loud threats and recriminations are interchanged.

There are clubs of innumerable designs, some comparatively light for the chase, and some very heavy, for hand-to-hand encounter.   These latter have sometimes rows of prominences carved upon them at the thick end to increase the severity of the blow.   The club tapers to both ends, which terminate in sharp points.   Wooden swords, to be wielded with one or both hands, are common, and shields both light and heavy, broad and narrow, the shield-handle being generally formed by scooping out a horizontal groove on the back, and leaving a

* Mr. Smyth quotes Mr. Ferguson on the antiquity of the boomerang. His evidence is, I think, conclusive as to the use of a returning weapon like the boomerang among the Aryan races of Europe at the earliest historical times.   ("The Aborigines of Victoria," vol. ii. p. 325)

short longitudinal bar intact on the solid surface. ' Hilaman ' or
' elimang,' now a common name for shield among the European
population, originally designated a small bark shield. The
wooden weapons are usually more or less carved, and are often
partially coloured, either red or white.

The stone tools comprise hatchets, chisels, and knives. The
tomahawk is shaped like a rude American axe, and is of all
gradations of sizes, from what might be used by a child to a
heavy stone head some twelve or fourteen inches long. The
most common material is a bluish-green stone which takes a
fine polish and has a clayey fracture. Axes made of stone so
much alike that a superficial glance could detect no difference
in appearance, may be found in places a thousand miles apart.*
I have a broken axe-head which I found on the beach at Portarlington, no doubt the remains of an object that must have
been greatly prized. The dark blue ground of the stone is
starred over with milk-white specks. It is beautifully polished,
and back farther from the edge than usual. The axes were
ground to a cutting edge of crescent outline. The axe-handle
in some makes tapered almost to a point at the end to be held
in the hand.   It was made either of a tough vine or of a split
sapling of suitable thickness. The piece of vine or wood was
doubled.   In the loop thus formed, the head was balanced and
secured with cord and resin on the side next the haft.

The chisels had sometimes handles of bark wrapped round
them.   Besides the tools already mentioned there were stones
for pounding food, whetstones, shells for dressing weapons,
bone awls; twine made of wood-fibre, sinews or fur; fish-hooks,
nets, fishing-lines; water-vessels, such as koolimans (made of a
hollow knot of a tree or from the bend of a limb), calabashes,
and even human skulls; the appliance already described as a
climbing-rope; and various other local or less important implements.

Except in the case of particular persons or on particular
occasions, hardly a living thing was rejected as an article of diet,

* Mr. O. R. Rule, of the Technological Museum, Melbourne, has favoured
me with the precise names of the stone of four axes. One from the Burnett
District, Queensland, and another from the Upper Darling, N.S.W., are
aphanite greenstone; a third found at Cheltenham, Victoria, is diorite; the
fourth, mentioned above as found at Portarlington, Victoria, is diabase
porphyry.

from the cicada to the kangaroo. The black man's table was
thus furnished with animal food of all kinds and flavours.
Grubs found in green trees were highly esteemed; so were
snakes, bandicoots, porcupines, emus, and men. When hungry,
flesh would be eaten raw with avidity, but if time permitted it
was roasted. A common practice was to bite off portions as
they were cooked, the joint being handed round for each
member of the group to take a bite, and then placed on the fire
again. Honey, the product of the native bee, a very tiny, innocuous, slow insect, was very much in request where it was
obtainable. In Queensland there was an ingenious and convenient way of eating honey which may possibly have been
practised elsewhere as well. A sheet of the inner, tough,
fibrous bark of a tree was procured. This was rubbed and
softened until it became like a piece of thin matting or old
bagging two or three feet square. It then formed a spongy
rag, and part of it would be dipped in the honey and afterwards
sucked by one after another of the members of the family from
the head of the house downwards. Even when the honey in
substance had got exhausted the flavour would cling to the bark
for a long time, and would reward the sucker for his exertions,
and form a treat to offer a friend. It was certainly a very
social form of enjoyment, and an economical mode of taking
food; whether the reader would care to join in it is another
question.

The supply of vegetable food was much more restricted. A
kind of grass-seed called ' nardu ' was used by the natives in
the north-west of New South Wales. This is different from the
' nardoo ' of Central Australia, now familiarly known as the food
which Burke, Wills and King tried to support themselves upon
at Cooper's Creek. Fern roots and the Australian yam, a
species of Dioscorea,* are perhaps the most common edible
vegetables. Other kinds, whether the roots, stems, or fruits be
eaten, are local products, different districts producing food
peculiar to them. The zamia nut is eaten within the tropics,
certainly in the west, and probably in the east also. In the
south of Queensland a plant like the cassava or arrowroot
grows on the banks of streams, and its root is eaten when
pounded and freed from the juice, which is excessively pungent.

* Grey's " Journals of Expeditions of Discovery," vol. II. p. 12.

The same locality is distinguished for the beautiful Bunya-tree, the *Bidwillii Araucaria*, an ornament of the scrubs on the high lands. The cone of this tree is of gigantic size, and in each scale there is an eatable ovule, which when mature is an inch or an inch and a half long, and about half an inch thick. The ovules are of conical shape, like an almond kernel, and covered with a tough envelope. When tender the fleshy part is all eaten. As the seed matures and the embryo assumes a definite shape, the surrounding tissue is drier and less palatable, and the embryo is rejected. When matured the natives prefer to eat the bunyas roasted. The kernels are also pounded into a kind of meal called 'nangu.' The bunya is a wholesome and much relished food. Individuals claimed special favourite trees as their own, but generally everybody had the range of the whole forest. The boles are often from two to three feet thick, perfectly straight and without a branch for the first fifty or a hundred feet, above which the branches spread into a beautiful dome-shaped top. The climbing-rope is called into requisition for the ascent, which is a difficult process, as the bark is flaky and jagged and the leaves are prickly pointed. The matured cones, as large as pumpkins, fall to the ground with a tremendous thud, on which occasions provision is had by picking it off the ground. About the same neighbourhood, and probably elsewhere if obtainable, the core of the top of a sort of cabbage palm forms a very juicy palatable food. The 'nardu' grass seed of New South Wales has been mentioned above; it is pounded and eaten without separating the husk.

The plant known as *pitcheri* or *pityuri*, which grows in the interior, is very much esteemed by the natives for its stimulating property. It is first chewed, and then mixed with wood-ashes and the leaf of a plant known as *kombari*. Then, after baking, the preparation is complete, and it is carried about for use. It is said to have the effect of sustaining the strength under severe exertion without any other food. The natives now chew it like tobacco, and take turns at the same quid.[*]

Along the marshy grounds of the Murrumbidgee and Lachlan Rivers a plant grows profusely which is locally known as 'combungie' or 'wangle.' The plants attain a height of seven or eight feet. They have a tap-root a foot or eighteen

* Curr's " The Australian Race," vol. ii. p. 38.

inches in length.   These roots used to be pulled up and collected by the women of a small community.   An excavation of circular outline was made in the ground, averaging three to four feet deep and fifteen to twenty feet across.   Half a ton of roots might be gathered for a large oven and placed in the centre on a great pile of dry wood.   On the surface were strewn layers of long grass and light sticks.   Then the combustibles were kindled and the excavated earth returned as a covering.   The time required for cooking depended upon the size of the oven, and might be several days.   When the 'wangles' were thoroughly done, water was continuously baled on to the oven until the whole mass was cooled.   It was then opened and the food came out almost white as snow and not unlike parsnips or potatoes cooked.*

This wholesale culinary operation was conducted much after the style of meat-roasting by the ovens that are so numerous in Victoria, where I have seen the *débris* or middens over twenty feet in diameter, with a corresponding height, the slope of the sides being rather less than the angle of deposition, and the top flattened by obvious causes.   In Victoria the ovens were used in the following way: A rude paving having been laid, a great quantity of stones and earth was heated by being heaped upon a huge fire of wood.   Then the fire was withdrawn, and the game, unskinned, was placed in the centre upon a layer of grass, more grass being strewed over it.   The heated stones and earth were next piled on top and the oven was left thus until the meat was cooked, which would then be taken out and the skin would easily peel off.

The diseases to which the aborigines are specially subject are rheumatism and pulmonary complaints.   These, though aggravated by changed habits since contact with the whites, are probably no new troubles.   Syphilis, introduced by Europeans, has terribly debilitated the constitution and corrupted the blood, but the scourge which sweeps off most of the natives is consumption.   Indigestion and toothache are common, dropsy and heart disease also occur.   All sickness from internal, unknown causes was attributed to sorcery practised by an enemy.   They possessed little or no knowledge of medicine, any remedies being almost exclusively externally applied.   A common treat-

* My informant is Mr. Humphry Davy, Balranald.

ment was for the doctor or sacred man of the tribe to suck the part affected and pretend to extract from it a pebble of the sort used as charms. There seems to be efficacy in the sucking, for a friend of mine who was suffering severely from an inveterate, inflamed eye, allowed a black "doctor" to mouth the eyeball, and the result of the treatment was immediate relief and speedy cure. Sometimes the doctor would apply a sacred stone to the part that was aching and profess to extract the cause of pain. From the analogy of a similar practice in the New Hebrides this may have been originally a kind of exorcism.

Wounds were often plastered with clay. In the case of sores on the limbs, circulation would be checked by the fastening of a ligature above the sore part. Mange was frequently caught from the dogs. There was a disgusting monkey-like method of dealing with it which I have seen practised. One person, using a short pointed stick, would prick the pustules all over the body of the patient, who would be reclining in a convenient posture and enjoying the operation. For headache a band was fastened tightly round the temples. Besides common remedial measures, such as those mentioned, each community would have methods peculiar to itself.

There is considerable difficulty in determining the length of life of the blacks, the generation born after contact with white people being, on the whole, very short-lived. From numerous instances it would appear that former generations were fairly long-aged. Almost every small community would have in it two or three men or women over seventy years of age, and here and there some centenarians would be met with. The impaired constitutions of the present generation, their unhealthy habits arising from a combination of native with European modes of life, the ease with which many fall into vicious practices, preclude the possibility of many of them attaining to hoar hairs. It seems very probable that, in Victoria and New South Wales at least, there will not be a single pure aboriginal surviving fifty years hence.

CHAPTER VIII

GOVERNMENT, LAWS, INSTITUTIONS

Government, laws, institutions—Aboriginal bondage to tradition—
Tribal cohesion—Leadership—System of kinship and matrimonial
restrictions—Ganowanian classes—Blood-ties or marks of courtesy—
Dr. Fison on the Murdoo legend—Classes not the result of a conscious
reformatory effort—Promiscuous intercourse—Polyandry—Exogamy
—Stages of social development as marked by marriage—Australian
classes, group-marriage—Negatives as names of communities, class-
names and totemism.

VARIOUS writers have shown that the noble savage is not the
child of liberty which he is popularly supposed to be. On the
contrary, while roving the forest in apparent security and
freedom his life is very uncertain, and from his childhood he
is shackled with burdensome ordinances inherited from his
ancestors, for the observance of which he usually has no intelligent reason to offer. The rules which prescribe the conduct of
the Australian aborigines are in every place numerous and
strictly obligatory, infraction being followed by penalties which
always involve the risk of injury to the person and often the
forfeiture of life. The unquestioning obedience which commonly
marks submission to these vexatious regulations is very striking.
The cohesion of a community depends entirely upon consanguinity and derives no strength at all from governmental
authority. A community is simply an aggregation of families
among which the older men have a certain amount of control,
derived naturally from age and experience. There is no recognised head, whether king or chief,* neither is there any definite
ruling body, elective or hereditary. Men of preponderating
influence are those who are distinguished for courage, strength,

* Some writers have recognised a distinct chieftainship, as for instance
Mr. James Dawson in the tribes he describes living in the southern watershed
of Victoria ("Australian Aborigines," p. 5).

and force of character. These, in conjunction with the elders, generally advise as to the public actions of the community, settle internal disputes, and enforce obedience to traditional law.   It is an abuse of language to designate the most influential man by the name of chief or king, as has occasionally been done, and an unwarrantable importation of foreign ideas into descriptions of Australian life./ People, speaking virtually one and the same dialect, will be/spread over from five thousand to ten thousand square miles of territory and sometimes more, and cut up into several small communities which, though usually friendly, may be involved in hostilities. Such a group of related septs would form what Mr. E. M. Curr has designated " associated tribes," association, however, being not entirely dependent upon close approximation of language. ' As a general rule, dissimilarity of speech connotes mutual internecine enmity, every stranger that falls into one's power being a proper object of slaughter. The so-called associated tribes barter with one another, inter-marry, and unite against a common foe.

/ To one accustomed to think only of the relations in civilised society, perhaps the most singular and conspicuous feature in Australian social life is the system /of kinship and the corresponding matrimonial restrictions. This point of study is particularly interesting and instructive, as bringing us face to face at the present day with a condition of society and inter-sexual relations which, from numerous instances existing in parts of the world widely separated, are generally believed to have universally prevailed at a prehistoric period, in what are now the most advanced races. Sir George Grey gets the credit of having been the first to place on record the Australian peculiarities of kinship and descent. While innumerable modifications are current, there are a few broad characteristics which mark the system and its accompaniments almost everywhere :

First.—A, being a male, his brother's children are spoken of as his own children, his sister's children are his nephews and nieces, his sister's grandchildren as well as his brother's are spoken of as his grandchildren; and if A be a female, with the interchange of the terms " brother's " and " sister's " the proposition is also true.

Secondly.—Every community is constituted by two or more

classes, most commonly four, and every individual bears one or other of the class-names.

Thirdly.—Descent is usually through the females, and this is especially marked by the class-name of the mother determining the class-name of the child.

Fourthly.—Marriage within the class is forbidden on pain of death; there is consequently exogamy in respect of classes, and usually tribal septs or communities are exogamous as well.

Systems of relationship like the Australian have been named by Mr. Morgan "classificatory."   Beginning with an examination of a form called the Ganowanian, prevailing among the North American Indians, he made a comparison of other forms in various parts of the world, and came to the conclusion that the names of relationships which at a first glance appear loosely and inappropriately applied, are names of blood-ties, and indicate communal marriage, or group allied to group at a more primitive time.   He is vigorously opposed by Mr. McLennan, who regards the relationship of the classificatory system as simply "comprising a code of courtesies and ceremonial addresses in social intercourse."[*]

The discussion of the merits of these two hypotheses would require a special monograph.   The writer inclines to the opinion that the terms used in the Australian system of kinship denote what were once blood-ties, and that their application was extended by analogy.   It does not follow that they are evidential of former group-marriage, unless the connubium of some own brothers with own sisters be understood by that name.   What seems to have originated such a theory is the fact that exogamous groups or classes are comprehended within one community.   The question of the former prevalence of so-called group-marriage will be settled by an accurate account of the origin of these classes, phratries or gentes, as they may be variously called.   To explain their origin, Mr. Morgan assumes that, following upon primeval promiscuous intercourse, there was marriage between a set of brothers and a set of sisters, and that a recognition of the resulting evils led society to deliberately partition itself into intermarriageable classes with a view to their prevention.

This field of inquiry so far as Australia is concerned has

* McLennan's "Studies in Ancient History," p. 273.

been ably and comprehensively exploited by two distinguished co-workers, Mr. A. W. Howitt and Rev. Dr. L. Fison. When writing on this subject in a paper contributed to the Royal Society of New South Wales in 1889 I felt constrained to dissent from some of their conclusions, and while I gladly acknowledge my indebtedness to their writings I am still unable to go the whole length with them. Dr. Fison * says they "have found it advisable to drop the term 'communal marriage' altogether because of its misleading tendency and to substitute 'group-marriage' for it." But while discarding an objectionable name they still adhere to the hypothesis for which it stood. Can the hereditary relationship subsisting between members of two intermarrying classes be properly designated by the name of marriage? I think not. It would involve in certain of the Australian class-systems the conclusion that a man was naturally and at the same time the husband of his recognised wife, his daughter and his mother-in-law. Dr. Fison says "the word marriage itself has to be taken in a certain modified sense," "what it implies is a marital right or rather a marital qualification." A right and a qualification are very far from equivalent. The latter term is appropriate, the former doubtful. An argument in favour of group-marriage based upon the application of terms designating real relationship to all the members of a group where there are four or more groups is met by the objection that in Australia the manifold groups have been derived from an original pair.

The Rev. Dr. L. Fison, in the work "Kamilroi and Kurnai," emphasises and corroborates Mr. Morgan's view. Dr. Fison, in dealing with the rise of the Australian exogamous classes, lays stress upon the Murdoo legend, an aboriginal tradition, the substance of which is that the classes restricting . matrimony were constituted to remedy the bad results of in-cestuous marriage. That these classes do prevent certain close marriages is true, but is it logical to conclude therefrom that they were inaugurated for this purpose? Moreover they require to be supplemented by other restrictions to prevent alliances between persons near of kin. The class-barriers would allow inter-marriage of cousins, and in some of the class-systems

---

* "Proceedings of the Australasian Association for the Advancement of Science," 1892, p. 689.

those unnatural alliances indicated above. It seems to me that the Murdoo legend is too flimsy to support such a conclusion, and *if the classes were due to some other cause than a conscious reformatory effort, their effects would still be the same.* I prefer to regard them as springing from natural conditions of life, having a reformatory tendency no doubt, but the reformation neither recognised nor designed by those who were the subjects of it.[*] The obstacles presented to intermarriage of persons near of kin have put inquirers upon what is probably a wrong scent. Independently I arrived at the view which Mr. McLennan takes, viz., that the matrimonial classes are memorials and results of the coalescence of different stocks of people, which were once distinct and exogamous tribes or races, and this view is in harmony with the theory of the origin of the Australian people enunciated in this treatise. Both Mr. Morgan and Mr. McLennan set society in motion under a condition of promiscuous intercourse. This is quite an imaginary starting-point, and reduces mankind to a state of degradation lower than the brutes, which in many cases, and especially in the case of the higher apes, go in pairs. With a view to accounting for the change of kinship through females to kinship through males, Mr. McLennan finds it expedient to make polyandry follow promiscuity, and the necessity for polyandry he finds in the infanticide of female children and the consequent disturbance of the balance of the sexes. But the prevalence of infanticide of female infants is only postulated, not proved; and although in various countries polyandry has been the rule, and in others has been practised to some extent, nevertheless, a polyandrous stage of society in all races is far from established. Judging from the propensities of humanity as witnessed at the present day in savage races, polygyny is a much more favoured form of connubium than polyandry. And

---

[*] " You will find by reference to Huth's 'Marriage of Near Kin' that the injurious effects of close intermarrying is a myth and hence cannot be the basis of the Australian horror of blood alliances." Mr. S. E. Peal, "Australian Association for the Advancement of Science," vol. v. p. 514. *Cf.* also Huth's "Marriage of Near Kin," pp. 138 and 353. Mr. W. E. Roth pronounces emphatically against the Australian classes having been formed to avert consanguineous marriages ; *see* Roth's "Ethnological Studies Among the North-West-Central Queensland Aborigines," p. 69. Westermarck also rejects Morgan's theory ; *see* his "History of Human Marriage," pp. 318-319, 544.

far more may be said in support of there having been as a rule a surplus of females in a community than the contrary. Nature herself tries to maintain the balance of the sexes and compensate for the greater mortality among males by an excess of male births. On the occasion of nations meeting in battle the victorious side slaughters the males and usually preserves the females, and then either for the conquerors or the conquered polyandry would seem too unnatural to be dreamt of.

Polyandry and agnation are bound together in Mr. McLennan's theory by rights of succession to inheritance; in other words, by property; and, moreover, the conditions of life with which he mainly deals to explain succession through males are those of semi-civilised peoples among whom both sexes have accumulated property. There is a stage farther back than this exemplified in Australian aboriginal life, at which there is scarcely aught but territory to claim, and it is tribal rather than personal property; and as for the women, with exogamy in regular operation, they possess nothing beyond a few threads, nets, baskets, or the like, about the succession to which there is likely to be no quarrel. At such a stage woman possesses practically nothing but her name and her charms, while she herself is man's most precious property. It seems to me that the primitive idea of acquiring and holding woman as one's own property is at the root of connubial systems, and in the majority of cases would conduce to polygyny rather than to polyandry. Let it be assumed that in the rudest state of society men covet women to be their peculiar possession and the following results, which obtain in Australia, ensue. The matured males by dint of force, and the elderly men by the authority of age, contrive to provide themselves with a plurality of wives, while the younger men must of necessity remain single, unless they procure partners by capturing them from an adjoining tribe. Women would thus be in continual demand, and exogamy would be conducted first in a purely hostile and predatory manner and later by barter or agreement of some kind.

From being in a sense inevitable, exogamy would become the normal mode of marriage, and the result would be that the wives in a tribe would be of a different tribe from their husbands, and would have the name or totem of their own tribe clinging to them. With indefinite paternity and definite maternity the children would belong to the mother and be recognised as of her blood, whatever general or tribal name the foreign mothers bore would also be attached to their offspring, unless the latter received a new special name from their hybrid appearance. Thus in process of time a homogeneous tribe would become heterogeneous in blood and embrace two, if not more, intermarrying classes, and tend to endogamy as regards the tribe, exogamy still characterising the classes.

Exogamy would tend to succession through males, even while there was uterine inheritance of class-names, because the sons would remain on their father's ground while the daughters would pass to other tribal territory, at first by capture and empty handed. But marriage *within* a heterogeneous community once reached, and personal rights in property admitted, there might be inheritance either through males or females.

The number of classes in an Australian community may vary from *two*, as among the blacks at Mount Gambier, to ten,* as among the Kamilroi; but the most common number is four, and there is good reason for concluding that at first there were only two classes which have been multiplied by subdivision or more probably by communities amalgamating. Six of the Kamilroi classes are certainly subdivisions of the four larger ones, and these again were either subdivisions of the two primary ones or the result of the combination of two different tribes. Where there are four classes they fall into related pairs, marriage being prohibited between the sections in one pair as if they formed just one class. These larger divisions have been called for convenience phratries. The rule is for a class of one phratry to marry into a particular class of the other phratry, the resulting offspring bearing the name of the remaining class in the mother's phratry, but occasionally either section of one pair of classes could marry into either section of the other pair. This is the case with the Kabi tribe of southern Queensland. The Kabi community has four classes—Barang, Balkun, Bŏnda, and Dherwen. Marriage is prohibited within any of the classes. Barang may not marry with Balkun, nor Bŏnda with Dherwen, but either

---

* Among the Narrinyeri of South Australia there were eighteen divisions called clans by Mr. Taplin, but which were virtually classes like the above, as they served the same purpose.

Barang or Balkun may marry either Bŏnda or Dherwen. There is this peculiarity to be noted about the descent which is perhaps also a proof that the four classes are subdivisions of a primary two, that the class-name alternates from mother to offspring by a continual recurrence of the same pair of names; thus one line of descent will be Barang, Balkun, and the other Bŏnda, Dherwen, *ad infinitum.*

Without postulating a fission of two classes into four, the existence of the four may be assumed to be due to the coalition of two communities which had each already two classes. Dr. Fison suggests * this solution of the multiplication of classes from two to four, and the writer thinks that no better can be offered. In support of the theory of multiplication of classes by fusion of tribes having each two or more class-names, I would point out that the terms 'Bŏnda,' 'Dherwen' are no doubt the same as the Kamilroi words 'bundar,' *kangaroo* and 'dhina-wan,' *emu*, and that the other related pair of terms probably just mean the same, *i.e.*, 'Balkun,' *kangaroo*, and 'Barang,' *emu*. I have it from native authority that 'Barang' means *emu*, and that 'Balkun' means *native bear ;* but at the junction of the Thomson and Barcoo 'balcun' is the name for *kangaroo*, and further, 'Balkun' is displaced by 'Bandur' on the Brisbane River, which is probably a variant of Kamilroi 'bundar,' *kangaroo*. At the Hastings River in N. S. W. 'Bulkoing' means *red wallaby* and 'Bundarra' *black wallaby*. Thus each related pair of the Kabi terms would mean *kangaroo, emu.*

Cohabitation between members of the same class is held to be grossly criminal, and is in many instances punishable by death. The union of individuals belonging to classes that cannot lawfully intermarry is equally abominated. Even in cases of rape the class rules are respected. The profound regard which the blacks show for restrictions fettered upon them by tradition, and for which they can give no better reason than that such is the practice, points to a very powerful originating cause and a sanction derived from condign and bloodthirsty penalties. To me at least, it is incredible that the segmentations into exogamous classes could have been deliberately made by agreement to avoid the evils of incest, for these would not be easily recognisable by nomadic savages. It seems more harmonious with

* "Kamilroi and Kurnai," pp. 71, 72.

social development to suppose that the gentes arose in the following manner. The women of a tribe were highly prized and jealously guarded by their husbands, whatever the type of connubium may have been, and bachelors, who, by reason of youth or other disability, could not obtain wives of their own tribe (*i.c.*, what subsequently, when two or more tribes were fused, became their class), were obliged to obtain them by capture. The danger of tampering with the women of their own tribe made exogamy the rule in course of time. There may also be an auxiliary cause to exogamy among barbarians in what may be called an instinctive hankering after foreign women.* Some light may be thrown upon the matrimonial classifications by Hamor's proposal to Jacob,† "And make ye marriages with us and give your daughters unto us and take our daughters unto you and ye shall dwell with us." Had this overture not miscarried, two families might have amalgamated and become "one people" as was proposed, embracing two intermarrying but exogamous classes. In this instance the cross-marriages would have begun by compact not by capture, and subsequent historians or ethnologists might have accounted for the rise of the classes by a supernatural wisdom like that which characterises the Murdoo legend.

Messrs. Fison and Howitt obliterate the Australian individual in the distant past, regarding him as merged in his class. The class is an entity of which one person is only a fragment, and all the members of a class have marital rights over all the members of the class or classes with which they may intermarry. This is the hypothesis founded upon an incomplete induction from several practices now extant. It is impossible here to traverse the whole question, but having carefully weighed the arguments in favour of group-marriage, while admitting that there is a good deal in them to point to it, I fail to see that communal or group marriage has been proven to exist; on the contrary, the conclusion contains much more than there is in the premises.

---

* I am gratified to observe that Westermarck approves of the above remark as a recognition of a psychological fact, and that it suggests a reason for exogamy virtually the same as that which he has enunciated. "The History of Human Marriage," pp. 321 and 546.

† Gen. xxxiv. 9, *et seq.*

GENERAL VIEW OF AUSTRALIAN CLASS-SYSTEMS

M = Male; F = Female; C = Children

VICTORIA.

In centre and north two exogamous classes, Bunjil (eaglehawk) and Wah (crow), distributed so that natives born of one community are of one class.   C named after father.

In extreme east (Gippsland) sex totems prevailed.   All males, Yīrung (emu-wren), all females, Djītgun (superb warbler).

In south and west originally two classes, Kurokeetch and Kappatch (Dawson), more probably Kunmait, one of the primary classes.

Phratry I. { Kurokeetch (long-billed cockatoo), M marries into any class save those of his own phratry.
Kartpœrap (pelican) ,, ,, ,, ,, ,,
Phratry II. { Kappatch (Banksia cockatoo) ,, ,, ,, ,, ,,
Kirtuk (carpet-snake) ,, ,, ,, ,, ,,
Kunamait (quail) ,, ,, ,, save his own.

C are of mother's class.   The names take a feminine termination, thus : Kuroka heear, Kartpœrap heear, Kappa heear, Kirtuk heear, Kunamit heear.   The system extends into South Australia, where, at Mount Gambier, it is varied thus :

Kroki, M marries Kumaitgor.   C are, M, Kumait; F, Kumaitgor.
Kumait, M marries Krokigor.   C are, M, Kroki; F, Krokigor.

Each class has five sub-classes or 'boorts' under which all objects in nature are ranked.

NEW SOUTH WALES.

On the *Lower Murray* and *Lower Darling Rivers* two exogamous classes :

Kilpara (eaglehawk), M marries Mukwara (crow) F } C take mother's class name.
Mukwara (crow) „ „ Kilpara (eaglehawk) „ }

The system on the Upper Murray and Maneroo Tableland has same classes with change of names :

Merung (eaglehawk), M marries Yukembruk (crow), F ; C *belong to father's class.*

Covering most of New South Wales and extending into S.W. of Queensland there are four classes (originally two). The sub-classes vary in number in different localities :

Phratry I. Kupathin	1. { M. Murri F. Matha	{ 1 sub-class Duli (iguana) 2 „ Murriira (paddymelon)	M marries Matha. Murriira, or any Butha. „ „ „ Duli „
	2. { M. Kubbi F. Kubbotha	{ 3 „ Mute (opossum) 4 „ Murriira (paddymelon) 5 „ Duli (iguana)	„ „ Kubbotha, Duli, or Ippatha Dinoun. „ „ „ Mute „ Nurai. „ „ „ Murriira „ Bilba.
Phratry II. Dilbi	3. { M. Kumbo F. Butha	{ 6 „ Nurai (black snake) 7 „ Dinoun (emu)	„ „ Butha, Dinoun, or any Matha. „ „ „ Nural „
	4. { M. Ippai F. Ippatha	{ 8 „ Dinoun (emu) 9 „ Nurai (black snake) 10 „ Bilba (bandicoot)	„ „ Ippatha, Nurai, Kubbotha Duli, or Kubbotha Murriira. „ „ „ Dinoun, or Kubbotha Mute. „ „ „ Nurai „ Murriira.

Rules of Descent : (1) The second or totem name of C is the same as that of mother ; (2) C of a Matha are Kubbi and Kubbotha ; those of a Butha, Ippai and Ippatha ; those of an Ippatha, Kumbo and Butha ; those of a Kubbotha, Murri and Matha.

GENERAL VIEW OF AUSTRALIAN CLASS-SYSTEMS—*continued*

M = Male; F = Female; C = Children

QUEENSLAND.

In southern coastal districts four classes :

Phratry I.	{ Barang (emu)	M	marries either	}	Bunda	C Dherwen.
	Balkuin (kangaroo or red wallaby)	,,	,,		or Dherwen	C Bunda.
Phratry II.	{ Bunda (kangaroo or black wallaby)	,,	,,	}	Barang	C Balkuin.
	Dherwen (emu)	,,	,,		or Balkuin	C Barang.

Inland, west of Balonne R. :

Phratry I.	{ Urgilla	F	Urgillagun.
Malera	Unburri	,,	Unburrigun.
Phratry II.	{ Wunggo	,,	Wunggogun.
Wuthera	Obūr	,,	Obūrūgun.

In neighbourhood of Mackay coastal district the subjoined classes correspond to above :

Phratry I.	{ Gurgela	F	Gurgelan	Gurgela marries Kubaruan	C Wungo and Wungoan.
Yungaru	Bunbai	,,	Bunbaian	Bunbai ,, Woongoan	,, Kubaru ,, Kubaruan.
Phratry II.	{ Wungo	,,	Wungoan	Wungo ,, Bunbaian	,, Gurgela ,, Gurgelan.
Wutaru	Kubaru	,,	Kubaruan	Kubaru ,, Gurgelan	,, Bunbai ,, Bunbaian.

SOUTH AUSTRALIA.

On coast south-east of Adelaide, the Narrinyeri tribe.

Eighteen exogamous clans or local classes, each having its own totem.   C belong to father's clan.

Port Lincoln two classes—viz., Matteri and Karraru.   Classes exogamous.   C take mothers' class-name.

Near centre of Australia, Urabunna tribe west of Lake Eyre has two primary classes with four sub-classes in each, designated by totems.

I. Matthuri {
Cicada
Dingo
Emu
Swan

II. Kirarawa {
Crow
Waterhen
Rat
Pelican

Cicada, M marries Crow, F, and so on with parallel classes.   C take mother's totem name.

At the Macdonnell Ranges, Arunta Tribe.

Phratry I. {
Panunga M marries Purula   F   C are Bultharra.
Kumarra „   „   Bultharra   „   C „ Purula.

Phratry II. {
Bultharra „   „   Kumarra   „   C „ Panunga.
Purula   „   „   Panunga   „   C „ Kumarra.

The above are given by Prof. W. B. Spencer and Mr. F. J. Gillen.   Mr. Willshire's equivalents are Aponunga, Coomarra, Pultarra, Perula.   Further north the classes, Panunga, &c., are increased by four additions.   Panunga adds Uknarria, Kumarra adds Umbitchana, Bultharra adds Appungerta, and Purula adds Ungalla.   The added classes marry only among themselves, thus: Uknarria M marries Ungalla F.   Hence there results an imperfect amalgamation of two distinct class-systems.

GENERAL VIEW OF AUSTRALIAN CLASS-SYSTEMS—*continued*

M = Male; F = Female; C = Children.

WEST AUSTRALIA.

In the north-west at De Grey River and Nickol Bay the same system is in vogue as at Alice Springs, so that it extends westward in a direct line, perhaps with interruptions, a distance of 1200 miles.   The Nickol Bay classes are :

Phratry I.	Panaka	M marries	Booroongoo F		C	Palyeery.
	Kymurra	„	„	Palyeery	„	„ Booroongoo.
Phratry II.	Palyeery	„	„	Kymurra	„	„ Panaka.
	Booroongoo	„	„	Panaka,	„	„ Kymurra.

The above are given in Curr's "The Australian Race," vol. i. p. 298.   At De Grey River there is a slight variation in names: Banakoo, Kiamoona, Parrijari, Purungnoo.

In neighbourhood of New Norcia there is a system of six classes, divided into two primary classes, one class in each of the primary being restricted to marry out of its phratry, two classes in each primary having the right to marry into their own phratry :

Phratry I.	1. Tirarcp	may marry into	4, 5. 6.
	2. Mondorop	„	3, 4, 5, 6.
	3. Tondorop	„	2, 4, 5, 6.
Phratry II.	4. Palarop	„	1, 2, 3, 5.
	5. Noiognok	„	1, 2, 3. 4.
	6. Jiragiok	„	1, 2, 3.

In neighbourhood of Albany the two principal divisions are Munichmat (white cockatoo) and Wordong (crow).

NORTHERN QUEENSLAND.

In north-west the system as represented in the Pitta Pitta tribe is :

Phratry I. Ootaroo	{ Koopooroo	M marries	Koorkilla	F	C	Bunburi.
	{ Woongko	,,	,,	Bunburi	,,	,, Koorkilla.
Phratry II. Pakoota	{ Koorkilla	,,	,,	Koopooroo	,,	,, Woongko.
	{ Bunburi	,,	,,	Woongko	,,	,, Koopooroo.

Mr. W. E. Roth ("Ethnological Studies Among the North-West-Central Queensland Aborigines," pp. 56–70) gives the following comparison :

	Pitta Pitta	Kalkadoon	Miubbi	Workoboongo	Wollangama, Normanton	Purgoma, Palm Isles, N. of Townsville	Jonon sub-tribe, Cooktown
Ootaroo	{ Koopooroo	Patingo	Badingo	Patingo	Rara	} Naka	} Chepa
	{ Woongko	Kunggilungo	Jimmilingo	Jimmilingo	Ranya		
Pakoota	{ Koorkilla	Marinungo	Youingo	Kapoodunga	Awunga	} Tunna	} Junna
	{ Bunburi	Toonbeungo	Maringo	Maringo	Loora		

The correspondence of the Pitta Pitta terms with those used at Mackay and west of Balonne River is well marked.

Tribes at Cape Grafton, Mulgrave River, and Lower Barron have two exogamous classes, Gooroogooloo and Goorubenah. C take *father's class-name.*

Authority : Rev. E. R. Gribble in "Science" (Austr. Anthrop. Journal), March 1897, p. 84.

What Dr. Fison's facts go to show is the scrupulously fenced chastity of classes in respect of those classes with which they do not intermarry. As I have no desire to underrate the evidence for group-marriage, I may mention a practice said to have prevailed among the Kabi, which may be used to support the theory of group-marriage if it do not rather indicate the assertion of a right to share in a wife's favours by those who have helped to capture her. I refer to the occurrence of the *jus primæ noctis*, which I heard of first from the lips of a white man who had on occasions lowered himself to the level of the aborigines, and which was afterwards certified to me by a black boy. Mr. D. Campbell informs me that the elders of the tribe claim the same right in the South Gregory District. Mr. W. E. Roth * mentions the first night's promiscuity as regular in North-West-Central Queensland. It is also mentioned by Mr. F. Small † as attending marriage by capture at the Clarence River, New South Wales. But Dr. Fison seems to have overlooked that a blackfellow holds his wife as his own special property against all comers, and allows intercourse with her only as a favour or for hire. This is the rule, and jealousy though feeble in some aboriginal communities is well marked in others, and is stamped not only on custom by the violent beating of the unfaithful spouse, but on the language by a special name.

Prof. W. B. Spencer has courteously informed me that the researches of Mr. Gillen and himself in Central Australia have yielded results corroborative of Dr. Fison's views.‡ The forthcoming work by Prof. Spencer and his colleague will be as valuable as interesting, but whether it will place the group-marriage hypothesis beyond question remains to be seen.

The classes are most commonly designated by names of animals, especially eaglehawk and crow in the south-east; and emu, kangaroo, iguana, opossum, turtle, snake, native, bear, are common names elsewhere. In some parts the names

---

* "Ethnological Studies Among the North-West-Central Queensland Aborigines," p. 174.

† "Science" (Australian periodical), March 1898, p. 47.

‡ The special evidence is, I understand, exact information about the practice of what has long been known as *women being granted paramours, i.e.,* a woman being allied to a number of men at the same time who possess graduated preferential rights over her.

of plants and various other objects are also employed as class-names.

An inquiry into several peculiar usages is suggested here, viz., the mode of naming communities, the nomenclature of the classes and the occurrence of totemism. In New South Wales and Queensland especially, but not exclusively there, a community derives its own name and the name of its language from one of its verbal negatives. Unless a more reasonable ground for this style of designation can be adduced, the writer would be disposed to account for its origin in the frequent repetition of "No, no," by persons when addressed in a dialect which was not fully intelligible to them. It very frequently happens among ourselves that a man is nicknamed from a word which he is fond of using, and we need only to extend this mode of naming to a community having one speech to be able to give a rational account of the origin of naming tribes from their negatives. A confirmation of this theory is found in the names of the languages at Byron Bay, Richmond River and Tweed River, which are called respectively 'minyung,' *what;* 'nyung,' *what;* 'ngando,' *who.* One tribe, the Pikumbul, on the Dumaresque River, New South Wales, is named from its affirmative, the reason for the imposition of a name from a negative will suffice to explain the derivation of one from an affirmative, viz., excessive iteration of some word. Other tribes again are named after some animal, such as the eaglehawk—*e.g.*, the Meebin tribe, near Point Danger.

There must have been a time when all the Australian tribal names could have been counted on the fingers of one hand. What was their significance in that primeval day? It is hardly probable that they were derived from negative adverbs. It is more likely that they were names of animals, as appealing vividly to the imagination, the echoes of which we still hear in the eaglehawk and the crow of the south-east of the continent. If the original tribal names were names of animals, and if the gentes are monuments of distinct ancient races, the gentile names are at once accounted for. There must have been to the savage mind a valid reason for the adoption of such names: perhaps a fanciful resemblance between particular families and certain animals, perhaps an attempt to explain human origin on a development theory; at all events, the principle of nomen-

clature once adopted, its application could be indefinitely
extended, as it evidently was.   Judging from the recognisable
vestiges of this system of designation, it would appear to have
been in vogue in prehistoric times among the whole human
race.

The characterising of gentes or clans or tribes by animal
names is manifestly related to totemism, though not identical
with it.*   The animal the name of which is borne by the class,
is not usually venerated by the members of that class, in fact
the significance of the class-name is sometimes lost altogether.
Where totemism prevails—and it is pretty general in Australia
(though its existence may not be known to the whites of the
locality)—each individual in the tribe bears the name of an
animal or plant which is his totem.   His totem is revered and
protected by him, and although he may eat of the totems of
others he will not injure or eat of his own, unless compelled by
starvation to do so.   Natives of the Narrinyeri tribe do not
scruple to eat their totems.   Among them also the bearers of
the same totem constitute an exogamous clan.   At Mount
Gambier, Victoria, there are two exogamous classes, Kumait
and Kroki, each divided into five sub-classes† which bear
totems, and under the sub-classes all natural objects are classified.
In this case marriage is independent of the totem.   I believe
that totemism in a more or less pronounced form prevails
throughout Australia, even where not recognised by Europeans.
I remember seeing a black boy playing with a little lizard.   I
thought he was cruelly using it, and remonstrated.   He dis-
claimed hurtful intentions, and declared that it was a friend
of his; and another black boy confirmed his statement.   I
did not know at the time the importance of this admission,
or I would have followed up the discovery by inquiry, but
I am of opinion that this was a trace of totemism, the existence
of which in the tribe referred to none of the whites had any
idea of.

* The article on Totemism in the "Encyclopædia Britannica" includes
naming of tribes after animals in the system of totemism.   Whether this
should be done or no is simply dependent upon the wider or narrower
definition of totemism.   The definition in the article requires "superstitious
respect" for the animal after which the group is named.

† Curr's "The Australian Race," vol. iii. p. 461.

It seems probable that the clan-name and the totem were once identical, but that in certain places they have become differentiated and the application of the principle of naming after animals has become extended. By the Narrinyeri a man's totem is called his 'ngaitye.' The Rev. G. Taplin refers * to a statement made by Dr. G. Turner about a form of Samoan fetichism closely resembling the Australian totemism. A man's god may appear in the form of some particular animal, which thenceforth becomes his object of worship and is protected by him, and the name for such animals is 'aitu,' *i.e.,* gods.

Whatever may be the local peculiarities of totemism, its world-wide occurrence proves that it has been inherited from the common ancestry of the now much differentiated peoples who retain it, and that therefore it is almost as old as Adam, and part of the baby-clothes of the human race.

Prof. W. B. Spencer and Mr. Gillen have brought to light certain most interesting particulars regarding the totemism of the Arunta tribe of central Australia—notably, (1)† that totems are attached to localities, the totem of a child being determined by the place at which it was conceived. The reason given for this is that in the Alcheringa (a mythical period) one of the beast-man ancestors died at that spot; his spirit still dwells there, and enters into such women as conceive there, coming to life anew in the child; the tree or rock which the spirit-child is supposed to have inhabited before conception is called its ' nanja' tree or rock. (2)‡ That the imitation of animals at the initiation ceremonies is the representation by individuals of the actions of their particular totems and, at the same time, "each performer represents an ancestral individual who lived in the Alcheringa. He was a member of a group of individuals, all of whom, just like himself, were the direct descendants or transformations of the animals, the names of which they respectively bear. It is as a re-incarnation of the never dying spirit, part of one of these semi-animal ancestors, that every member of the tribe is born, he or she bears of necessity the

* "Native Tribes of South Australia," p. 64.
† "Proceedings of the Royal Society of Victoria," 1897, pp. 24, 25.
‡ *Ibid.* pp. 153, 154.

name of the animal, or plant, of which the Alcheringa ancestor was a transformation or descendant."

We may reasonably conclude that the general name of these totems or spirits, 'ngaitye,' 'aitu,' and 'nanja' * (New Hebridean 'ata,' 'nata,' *person, soul, spirit*) are radically the same and constitute a bond of relation between the Australian and Pacific Islands superstitions.

* 'Nanja' is strictly the haunt of the totem or spirit.

CHAPTER IX

MARRIAGE, MAN-MAKING, MUTILATIONS, BURIAL CUSTOMS

REVERENCE for age and authority has greatly aided the elderly
men in monopolising the wives of the class with which they
intermarry. Betrothals are exceedingly common, a female child
being usually betrothed by her guardians to some elderly friend
who attaches her to his household when she is perhaps not more
than twelve years of age. Elderly men have been seen actually
nursing children their own prospective wives. Betrothal is, I
think, founded on barter, the father or brother or father's
brother having the right to give the maiden away. Brothers
thus betroth their sisters in exchange for women to be their
own wives. Side by side with the betrothal system is that of
elopement, which nowadays is usually more fictitious than real.
It is in cases of elopement that the guardians of the female
demand satisfaction from the man with whom she has levanted.
A tremendous tempest of wrath is feigned, and no doubt the
combat is not unattended with risk, but after it is over the
cloud of anger and ill-will is completely dissipated. There are,
besides, instances of real elopement, after which the woman, if
caught, will be severely handled, and the paramour will receive
a sound thrashing in real earnest if the injured person be
powerful enough to administer it.

Marriage by capture takes place between members of hostile
communities. Sometimes a surprise party will be organised to
attack a camp, slaughter the males and abduct and appropriate

the females. This wholesale abduction is paralleled by individual cases of forcible abduction, on which occasions the women, if resisting, will be cruelly beaten.

In Gippsland marriage by agreement is the rule, pursuit and capture being feigned.* With varying details, marriage by mutual consent will be found in other parts of Australia as well, but not reaching consummation exempt from the results of marriage by elopement. The use of love-letters is perfectly understood by the Kabi natives of Queensland. The love-letter is a bit of a twig about an inch and a half in length, and marked with three small transverse notches, the middle one representing the 'dhomka' or postman, the other two the lovers. I have seen one of these in course of transmission. A black boy fished it out from the lining of his hat, where he had it sewed up. He carried it in this receptacle for several months until he had an opportunity of delivering it to the damsel for whom it was intended. The aboriginal pair had met and fallen in love at a great festive gathering some time previous, and the love-letter was a sort of expression of adhesion to engagement. These forms, therefore, of marriage occur: marriage by betrothal, by elopement, by forcible abduction, by capture, and by mutual consent, the practice varying with the community.

The mutual avoidance of mother-in-law and son-in-law may be conveniently referred to here. It is noticed throughout the continent and prevails in the South Sea Islands as well. One explanation which is offered for it is the abhorrence of incest, but this is not satisfactory, for if this were the reason there would be quite as strong grounds for shunning intercourse between mother and son, father and daughter, brother and sister. This last condition, the separation of own brothers and sisters, seems to be fulfilled in Fiji, but as it does not hold elsewhere it may be explicable on other grounds. The Rev. Dr. D. Macdonald supports with warmth the hypothesis that detestation of incest is at the root of mutual avoidance between a man and his relatives-in-law, and he gives some interesting facts about this practice in Oceania.† For instance, a husband has to shun his father-in-law as well as his mother-in-law, and all the females of the same *gens* as his wife. All these persons avoided bear in

---

* A. W Howitt, "Kamilroi and Kurnai."
† Rev. D. Macdonald's "Oceania," p. 181, *et seq.*

reference to the husband the same term of relationship.* Perhaps an etymological examination of that term might be of service. But if danger of incest be the ground for avoiding the mother-in-law it cannot be in the case of a man the reason also for separation from his father-in-law, and the mutual avoidance of these relatives is required in at least one part of Australia as well as in Oceania.† Moreover, here is a very pertinent question· to put to the facts surrounding this peculiar restriction, why is it that the daughter-in-law is not tabooed in the same way as the son-in-law? There appears to be no danger of incest in her case. It seems to me that the cause of estrangement is that the son-in-law has been in times long past guilty of an offence which his wife's relatives, and especially her mother, grievously reprehend, and which custom forbids the latter to condone, and the offence, it is most natural to conclude, has been the forcible abduction of his wife. Dr. Macdonald cannot bear to think of such a brutal state of things being normally tolerated, even among barbarians, but our moral sensitiveness should not blind us to the testimony of facts, and we know that marriage by capture was not uncommon in recent times even in the Highlands of Scotland, a notable instance in the last century being the second marriage of the mother of the celebrated Flora Macdonald.‡ The Rev. Dr. L. Fison mentions the fictitious concealment of certain persons after the death of a Fijian chief. In one place the henchman of the chief keeps out of sight for a number of days after his master's burial. " He is supposed to vanish as completely as if he had been actually buried with the chief (which probably was once the practice), and if any of the tribes-folk should happen to meet him he is invisible to them." The same shunning of observation and intercourse takes place on the part of the two head under-

* Among the Kabi people of Southern Queensland, ' nulang ' means son-in-law, ' nulanggan ' mother-in-law, ' -gan ' or ' gun ' is the feminine termination, so that ' nulang ' designates the relationship on both sides. In Victoria, a word practically identical—namely, ' nalum ' or ' ngulum '—signifies the same relationship. The etymology of the aboriginal term is very desirable, as likely to throw light on this obscure subject. I suspect it to be connected with Malay ' kulawarga,' *relationship by blood.*

† Mr. D. Stewart in Curr's " The Australian Race," vol. iii. p. 461.

‡ " Flora Macdonald: Her Life and Adventures," by her Granddaughter, p 20. For evidence relating to the practice of capturing wives, the reader is referred to McLennan's " Studies in Ancient History," pp. 31-49.

takers at Vunda, after their chief's burial, their term of enforced retirement being nominally a year. "They paint themselves black from head to foot and never take their walks abroad until after dark. If compelled to go outside during the daytime they cover themselves with a mat and nobody takes the slightest notice of them; in fact, nobody is supposed to be able to see them. The fiction is kept up that they are invisible or rather non-existent." * The point of interest in these examples is, that people who have become obnoxious to their kinsmen are regarded as out of the way, and if seen are not *perceived*, the overlooking being suggestive of a former obligation to take satisfaction. Surely some such obligation as this explains the repugnance which a wife's friends are fictitiously regarded as bearing to her husband more satisfactorily than does the abomination of incest.

It is interesting to note that the New Hebrideans allege that the reason for a man and his wife's relatives keeping apart is that if they touched each other they would "become poor." † Dr. Macdonald thinks that the poverty would be originally supposed to be the curse of heaven upon incest. The explanation is conjectural. It appears that the Efatese wife is purchased from her parents, and that after the death of her husband she may be disposed of by his friends, but not returned to her parents until they refund the price that was paid for her. Thus the present facts of the transaction do not suffice to account for keeping apart from the dread of contact causing poverty, and the reason for separation must be looked for in customs antecedent to those now in vogue.

Subjection to certain rites marks transition periods in the life of the young. Among some tribes there is a series of practices to be complied with by the youths, beginning when they are seven years of age, and ending with their full initiation into manhood. The man-making is a universal Australian observance, and is attended with more or less ritual and severity, according to the tribe. The initiation to manhood was the occasion of immense gatherings to a particular sacred spot. There was commonly a large circle‡ made, with not infre-

---

* Fison's "Burial Customs of Fiji," *Centennial Magazine*, February 1889.
† Rev. Dr. D. Macdonald's "Oceania," p. 182.
‡ The ceremony was called by the Kabi people *dhur*, which means a circle;

quently a gigantic human figure sketched upon the ground within the circle ; but there was besides a secret place adjoining, where the more important and solemn part of the ceremony was conducted.   The natives had the greatest reluctance to admit Europeans to witness the proceedings on these occasions, and if by chance one should be present at the large circle, he would usually be absolutely forbidden to approach the more secret place.   Almost every tribe had details in the man-making ceremonies peculiar to itself.   The neophyte was generally required to keep serious and still, all levity being strictly prohibited ; he was sometimes obliged to fast, and various devices were employed to test his courage.   The initiators were of a different tribe or family from those to be initiated.   Fires were brought into requisition.   A great smoke was raised by burning green leaves, and the novitiates were permitted, if not enjoined, to view women at a distance through the smoke.   Some fires had to be stamped out by the youths with their naked feet. The young men were tempted to break the prescribed fasts by offers of food, to lose their gravity by comic representations, to exhibit fear by being subjected to treatment which would naturally excite fear.   The severest punishment was threatened for failure to undergo the tests.   After the ordeal had been successfully submitted to, the youth was eligible for marriage. At the rite of initiation a chip of wood like the toy bull-roarer was called into requisition, as were also the sacred pebbles.

The stages of initiation have been called by English writers *degrees*.   At each stage the neophyte receives a new title, and after the final he enjoys all the rights and privileges of full manhood.   The ceremonies embrace throwing up, plucking out the hair of the head in handfuls, head-biting by the initiators, evulsion of one or more teeth, cicatrising, spurting of human blood on the neophyte from incisions on others, circumcision, introcision or subincision, fire-treading, and sitting upon green leaves heaped upon a smouldering fire.   These ordeals correspond with practices in the South Sea Islands.   Even this year (1898) the fire ceremony was witnessed by Drs. Hocken

and also *kiwar-yengga*, man-making. The ceremony first became known to Europeans by the name *bora*. Rev. W. Ridley derives this from *bora*, a belt. Possibly correct, but from analogy I suspect a connection with Wiraidhuri. *burbang*, a circle also the initiating ceremony.

and Colquhou, of Dunedin, at Ubenga, an islet twenty miles south of Suva, and is described in the *Melbourne Argus* of May 24.

The primary objects were evidently to enforce self-restraint and to try courage. In addition to these, Mr. A. W. Howitt,[*] one of the few Europeans who have been privileged with a sight of the Bora ceremonies, affirms that instruction was given to the youths in religion and broad moral principles. The absence of reference to religious and ethical teaching in other accounts published raise a doubt as to its being usually given.

Mr. Geo. W. Anderson, of Cowwah, Woodford, Queensland, has supplied me with an interesting description obtained from a blackfellow, which, as it has never been published before, I give here. For brevity's sake I condense. I cannot vouch for precision as regards the relative time and order of the ceremonies. This purports to be the old practice of the Kabi and allied tribes of the Brisbane and Mary Rivers, and differs in certain particulars from an account which I received from a black boy, and which appears in my notes on the Kabi tribe in Curr's work, "The Australian Race."[†]

Various parties of blacks congregate at one spot, each party having several candidates for initiation. One party takes the boys out of one camp, the men there take boys out of the next, and so forth. The boys are never taken out for initiation by their own friends.

The boys to be initiated are placed within a circle; are left there all night without a fire, but are allowed opossum rugs, and are taken out in the morning. The rest of the blacks sleep some distance away. A large fire is made, which has to be stamped out by the boys jumping upon it. They are provided with the 'bunandaban' or bull-roarer, which they frequently whirl. The boys have to fast; their heads are wrapped up in opossum rugs, and they must not look up to the sky. The presence of women is strictly disallowed for a month. The novitiates are threatened with death if they laugh. The boys are brought by stages occupying several days from where the circle is to the main camp of the blacks, a man having charge of each boy. At sundown, when they have approached near

the main camp, a gin, who is painted (red), sings in the hearing
of the boys. This is the signal for the latter to approach the
big camp in a string, but they do not yet enter it. They still
keep a separate camp, round which they and their attendants,
who are painted, march four times. Towards daylight they
make a closer advance to the main camp, and in the evening
the procedure of the previous evening is repeated. A number
of fires are made in a line, upon which green leaves are placed
to cause smoke. The young men, beginning at one end of the
line of fires, jump upon one fire and clap their hands, and repeat
this process in fire after fire until they reach the end nearest to
·where the women are, and there they camp for the night. Two
or three attendants for each boy are now required. They are still
kept separate from the women and watched all day. A huge
fire is made, on which the boys have to jump until its extinc-
tion. Thereafter they are placed in charge of one man, and
the other initiators go to some distance and call out a name to
each of the novitiates in turn, these answering to their names.
The boys are then rushed forward by a number of their attendants and caught by others, who toss them up and let them fall
upon the ground. This is followed by a corroboree rendered by
the men. The faces of the boys are next covered up. They
are carried on the shoulders of the men, who pretend they are
taking them into a flooded creek. When they go back to the
camp they are let go. The boys are then threatened by a party
prepared to fight them. Hostile cooeys are given on each side.
Hilamans (*shields, Eng. plu.*) are painted for the combat. The
day following, the boys are compelled to fight their seniors.
There are other attempts made both to frighten the boys and
to make them laugh. Laughter is threatened with death. A
'manngur,' elsewhere 'koradgi,' *i.c.*, *doctor, sorcerer*, scars the
novitiate on the shoulder. When the cut is healed he may go
freely to the big camp, where no notice is taken of him; he
must still, however, camp apart. A gin, painted red and adorned
with a cockatoo feather in her hair, is brought to the boy's camp ;
the feather is transferred to the boy's head, and the gin retires
again. For several days he is not permitted to look round. At
length the gin goes near where the boy is; they touch each
other, and are thenceforth man and wife.

Mutilations of some kind or some cutting of the flesh are

everywhere practised. These vary with different tribes. Piercing
the septum of the nose is the most common practice of the kind.
Circumcision of the young men prevails in the central zone from
north to south. It was not observed by the aboriginal Papuans,
for it is unknown in the coast district of West Australia, in
Victoria, New South Wales, and the greater part of Queensland
from the coast inward. The absence of it, in fact, is a mark of
the indigenes. Of the inhabitants of Sumatra, Marsden says:[*]
" The boys are circumcised, where Mohammedanism prevails,
between the sixth and tenth year." In dealing with Australian
art and religion, positive demonstration will be given of Sumatran
intercourse and influence, and the perplexing question as to the
origin of circumcision in Australia is now, I think, satisfactorily
settled. Between Sumatra and Australia there is a clear water-
way, which would be easily and rapidly traversed during the
prevalence of the north-west trades. I am confident that circumcision has been introduced from that island, along with
certain religious or mythological ideas. This theory of its introduction explains its partial distribution and its entire absence
in the more purely Papuan parts, and seriously affects Mr. Curr's
ethnological division, for which one of the leading principles was
the presence or absence of this rite. This principle is manifestly
misleading. The distribution of this rite serves, however, to
show how either an invasion or an influence from the north
would most easily spread. All over Australia circumcision
would probably have prevailed in time but for British settlement.

The amputation of one or two joints of the little finger of
one hand is practised upon the young women of some tribes on
the Queensland coast,[†] a form of mutilation followed by some of
the Kanakas of the New Hebrides. At the Daly River, in the
Northern Territory, girls remove the first two joints of the right
forefinger by tying round the joint a thin skein of strong cobweb, which is left until the joint falls off.[‡] The knocking out
of front teeth—one, two, or four, according to the tribe—is an

---

[*] " History of Sumatra," p. 287.

[†] A practice similar to this once prevailed in Japan, but has been suppressed by Government, and survives now only in fiction. The extreme
phalanges of the third and fourth fingers of the right hand of a mother were
amputated before her daughter attained the age of twelve or thirteen.

[‡] Rev. D. McKillop's " Trans. Roy. Soc. South Australia."

old-world barbarity which has been perpetuated here. A similar custom is prevalent in Sumatra, where the women have their teeth rubbed or filed down. But the most horrible of all the mutilations is that which Mr. Sturt designated "the terrible rite." This bloody concision is done within the area where circumcision occurs, but is not so widely practised.* It is inflicted with a stone knife. To describe the operation in detail is outside the scope of this work, but we cannot avoid asking what object it is intended to serve. Mr. Roth† has satisfactorily demonstrated that it neither prevents coition nor procreation, and suggests that it has been adopted on the principle of imitation as a corresponding practice to forcible vaginal rupture. It seems more reasonable to regard the latter practice, which prevails coincidently with subincision, as a necessary consequence of this. I accept the view of Westermarck that the object is ornamentation and increased virility of appearance. This view is supported by the mode of circumcision followed in Tanna, New Hebrides. The Rev. William Gray‡ says that there the prepuce is so cut as to leave a wing on each side, forming a large lump underneath. "The longer the operation, the more a man does it make the boy." And Prof. W. B. Spencer § mentions that in central Australia, on the occasion of the rite of subincision being undergone, young men who have been operated upon once and even twice previously will voluntarily come forward and call upon the operators to enlarge the incision to the utmost. So that a pride is taken in the enlarged appearance. Where subincision is practised vaginal introcision becomes inevitable. No mutilation is more horribly cruel or disabling, but savages have little or no compunction with respect to their treatment of women.

The ornamentation of the body by cicatrices has already been referred to, and needs only to be mentioned here. What the writer has seen done has been solely for adornment, but it has been alleged that the pattern of the incisions serves in some

* Curr's "The Australian Race," vol. 1. p. 74. and also map.   "Finditur usque ad urethram a parte infera penis " (Eyre's description).

† "Ethnological Studies Among the North-West-Central Queensland Aborigines," p. 179.

‡ "Report of Australian Association for the Advancement of Science for 1892," p. 659.

§ " Proceedings of the Royal Society of Victoria," 1897, p. 171.

cases as a tribal or gentile badge (an allegation well confirmed), or as a mark of rank, * although what rank an aboriginal could claim I cannot conceive. If by rank is meant stage of initiation to manhood, the observation is correct. These incisions are another link relating the Australians both to the Polynesians and the people of India.

Nothing could exceed the dolefulness of the lamentations made for the dead. The crying is as much like the howling of the dingo as the wailing of human beings. It is carried on vigorously and persistently for weeks after the decease, and then broken off by occasional crying fits. Very commonly the corpse is flayed and certain portions of the flesh eaten. Some parts of the body will be preserved and carried about as relics or charms, such as the knee-caps, the shin-bone, the hand, the skin. In Gippsland the hand of a dead person is worn round the neck as a charm and as an instrument of sorcery, a practice similar to the preservation of the finger-nails (and portions of the fingers attached) of a deceased person by the New Hebrideans. Mothers will carry the dead bodies of their children on their march, even in a putrefying state. This, according to Mr. Curr, is also a kind of penalty inflicted upon young mothers who are blamed for causing their baby's death by carelessness. I am reluctantly disposed to doubt Mr. Curr's reason. I have other testimony of this and similar practices being followed purely from affection. The women especially cling affectionately to parts of the body of deceased relatives, a very creditable tenderness in those whose belief practically is that death ends all.

One mode of disposing of the dead is to bundle the bones into a hollow tree. I have found three or four tombs of this kind within an area of about four square miles. Before being thus disposed of, some tribes wrap the corpse in bark. A practice followed on the east coast of Queensland, and at a place so far distant as Encounter Bay, South Australia,† is to stretch the dead on an elevated platform of boughs until the corpse has become desiccated. A very general mode of burial is to prepare the body for interment by doubling the legs so that the knees

---

* Dr. Carroll, *Centennial Magazine*, October 1888. The present writer has personal testimony that tribes wear distinctive scars, but he has not been able to verify the statement.

† At Encounter Bay, after the flesh is decayed, the bones are burned.

will come under the chin;. the hands are then tied by the side, and the corpse is placed in a grave in this sitting position head upwards. I am informed that on the Lachlan and Murrumbidgee the dead body was deposited with the head towards the south.* In the north of Queensland cemeteries are to be seen where there are accumulations of skulls.

Immediately after a death the camp at night is resonant with hideous sounds. When first I heard the howls of despair it made my very flesh creep with horror, and to heighten the effect the mourners might be seen the greater part of the night hurrying hither and thither brandishing torches, with the object, it was said (I know not how credibly), of frightening away evil spirits. As might be expected, the grave is very shallow. I have seen one in the Burnett District, Queensland, with several short logs placed at the side of it on the surface, which are said by the blacks to represent the number of brothers the deceased had, and to indicate by their position relatively to the corpse the direction in which the brothers resided.

Unless when the cause of death is very obvious, such as a spear-wound, it is held to have been brought about secretly by another blackfellow. Diverse methods are adopted for the discovery of the murderer. For instance, among the Kamilroi an ancient shin-bone relic, wrapped in cord and some greasy matter, is held near a fire, and when it fizzles it is believed to point in the direction of the guilty person, who is then easily identified. In central Victoria a straw would sometimes be inserted in a small ant-hole or other perforation in the covering of the grave, and the direction in which the upper end would point would be the road to take to find the person who had caused the death. And then it might be the first blackfellow of another tribe who might be met that would be slaughtered in cold blood in revenge. Captain Grey testifies that, among the blacks of Western Australia, the dread of blind vengeance on the occasion of a death was extreme, because nothing could save an innocent person from being pounced upon either in obedience to some augury or for satisfaction of spite on the part of a sorcerer. The murderer had always to be sought for, and somebody would have to satisfy the demand. In many tribes the corpse is interrogated as to who was the cause of his death, and responses are obtained

* My informant is Mr. Humphry Davy.

generally by spells.   While in the act of lamentation for the
dead, the women would lacerate their bodies from head to foot
till blood would be streaming from innumerable small incisions.
The blood was allowed to dry upon the skin.   The fact that this
practice was forbidden to the Israelites shows its great antiquity.[*]
Near relatives of the deceased wore some token of mourning
upon the head, the usual practice being to attach tufts of emu's
feathers to locks of the hair, and leave them to drop off of themselves.   In some parts clay was plastered over a net upon the
head and allowed to harden until the whole assumed the form
of a skull-cap.[†]   After being worn for a time it was laid upon
the grave.   A form connected with mourning as practised by
the Murunuda, South Gregory District, was to cover the whole
body with lime.   Another custom in mourning was a prolonged
abstinence from certain kinds of animal food.   Mr. Bradshaw
informs me that at Ruby Creek, Kimberley, on the occasion of a
man's death his wives are clubbed to death with great ceremony
by the old married men.   This atrocity has not been noticed in
other localities.   We are not surprised to learn that while in
the same neighbourhood the corpses of men are wrapped up
in bark and laid on ledges in caves, those of women are
flung under bushes as if not worth attention.   My informant is
Mr. Froggatt.

* Deut. xiv. 1 ; Lev. xix. 28, &c.   The making bald is forbidden in the
same connection, which is also an Australian sign of mourning.
† Curr's " The Australian Race " vol. ii. p. 238.

CHAPTER X

ART, CORROBOREES

Art—Corroborees — Message-sticks of Malay introduction—Rock-paintings, where found—Mr. Giles' discovery at Lake Amadeus—Captain Stokes' discovery at Depuch Island—Mr. Norman Taylor's in Cape York Peninsula—Mr. Cunningham's at Clack's Island—Painting at Nardoo Creek, Queensland—Captain Flinders' discovery at Chasm Island—Captain Grey's at Glenelg River, N.-W. Australia—Authorship—DAIBAITAH—Mr. Bradshaw's discoveries at Prince Regent River described—Explained—Parvati—Siva—Mr. W. Froggatt's discoveries — NAURI — Hand-prints — Figures — Cave-paintings in New South Wales, and at Billiminah Creek, Victoria—Sample of work at Billiminah Creek—Rock-carvings near Sydney—The Australian muse —Corroborees.

THE skill shown in the manufacture of weapons has been already noticed. These were often ornamented with rude colouring and carving. Some of the carvings appear to me to be imitations of letters, and perhaps a careful examination of very old choice specimens may result in an interpretation of the hieroglyph-like characters. A throwing-stick figured in Mr. Smyth's work,* and spoken of by him in terms of warm approbation for its artistic merits, bears engravings very like the Sumatran letters, which will be referred to again below. I am strongly of opinion that the native message-sticks are imitations of the old Malay practice, prevailing at least in Sumatra, of writing upon bamboo and rattan canes. It is very natural to suppose that isolated Malays dwelling among the Australians would endeavour to correspond with each other in this way, agreeably to the custom of their native land, and that the Australian aborigines, observing that the characters were legible, would readily imitate the more intelligent race. A careful preservation of old message-sticks is desirable; perhaps some may

* Mr. R. Brough Smyth's "The Aborigines of Victoria," vol. i. p. 308, Fig. 88.

yet be discovered or may already be in our museums or in
private possession, bearing legible writing. Those now current
convey intelligence purely by sign-writing, not by alphabetical
characters, and require the bearer to interpret. The message-
sticks vary in length from 1½ in. to the length of a walking-
stick; the thickness is also variable, and the figuring consists of
pits, notches, strokes, curves and zigzag lines. The inner side
of the skins in opossum rugs was also scratched with rough
representations of a few common objects, generally drawn in
single lines.

The art of painting has been so little practised by the
aborigines of Australia, that to say they were ignorant of it
altogether would not be far from the truth. Some of them,
after contact with Europeans, have given evidence of considerable imitative power, but usually native pictorial art has not
risen higher than rude conventional sketches of men, kangaroos,
emus, turtles, snakes and weapons, done mostly in charcoal and
occasionally cut out on trees or graven on rocks. The linear
designs scratched on the inner surface of opossum rugs or carved
on weapons, and sometimes coloured red, black or yellow, are of
the simplest patterns. But at a few places, very widely apart,
specimens of art have been discovered immeasurably superior to
the ordinary aboriginal level. The only localities, so far as I
can learn, where this higher artistic skill has been exhibited,
are the following: Depuch Island, one of the Forestier group,
on the west coast of Australia, in latitude 20° 37′ S. and longitude 117° 41′ E.; Cape York Peninsula; Clack's Island, near
Cape Flinders, on the north-east coast of Queensland; Nardoo
Creek, Buckland's Tableland, Central Queensland; Chasm
Island, in the Gulf of Carpentaria; the Kimberley District,
Western Australia; a few other places in that quarter, and
especially the Glenelg and Prince Regent Rivers, not far inland,
on the north-west coast of Australia. Mr. J. Bradshaw informs
me that Lieutenant Oliver, of H.M.S. *Penguin*, while on a survey
expedition on the west coast of Australia, found cave-drawings
on Feint Island, near Bigge Point (latitude 14° 30′ S., longitude
125° 3′ E.), and took some sketches. I do not know their
character.

In three places, a few miles distant from each other, Mr.
Giles found paintings of inferior workmanship and accompanied

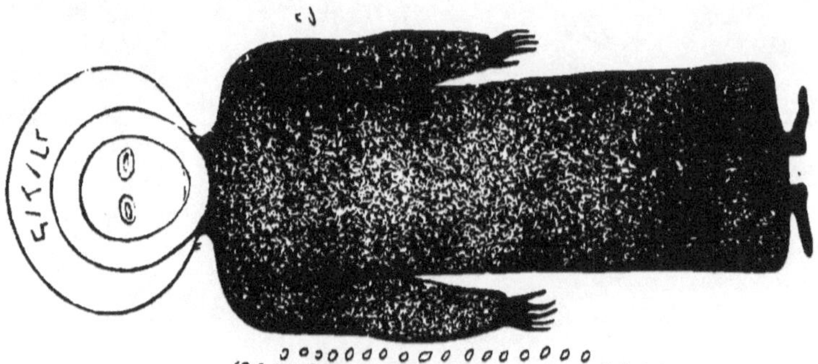

GREY

FIG. 2

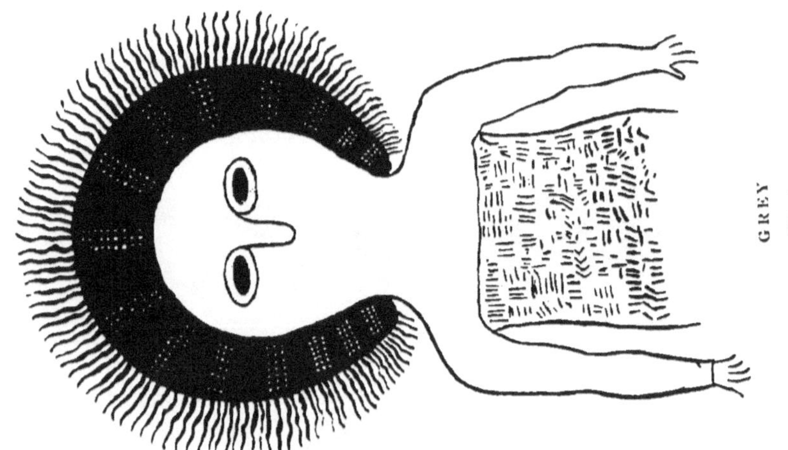

GREY

FIG. 1

by the almost universal hand-prints. He noticed characters like the Roman numerals VI painted red, and dotted over with spots. His discovery was made a little to the north of Lake Amadeus, near the heart of Australia, and the description he has given of the style of art suggests that the artists were of the same race as those who elsewhere have left such memorials of their presence.

The paintings on Depuch Island are numerous, but, judging from the sketches made by their discoverer, Captain Stokes, they are much inferior to the other groups in point of subject and treatment; they represent animals chiefly. In one sketch there is a rude attempt at delineating a corroboree. The artists have been satisfied if what they intended for human figures have been recognisable as such.

In the Cape York Peninsula, the northernmost part of Queensland, Mr. Norman Taylor, when exploring, "found a flat wall of rock on which numerous figures were drawn. They were outlined with red ochre and filled in with white. The figure of a man was shown in this manner, and was spotted with yellow." *

At Clack's Island, paintings were discovered by Mr. Cunningham, June 23, 1821, when he accompanied the King's Survey Expedition. "They were executed upon a ground of red ochre (rubbed on the black schistus), and were delineated by dots of a white argillaceous earth which had been worked up into a paste." They represented " tolerable figures of sharks, turtles," &c. Besides being outlined by the dots, " the figures were dotted all over with the same pigment, in dotted transverse belts "; † more than one hundred and fifty figures had been thus executed.

The work at Nardoo Creek, Queensland, must be very imposing if it be correctly interpreted. The picture is seventy feet across. It is said to represent a lake of fire, out of which are stretched life-size "dusky brown arms, in hundreds, in every conceivable position, the muscles knotted, and the hands grasping convulsively, some pointing a weird finger upwards, others clenched as if in the agonies of death." ‡

* Mr. R. Brough Smyth's " The Aborigines of Victoria." vol. i. p. 292.
† King's " Voyages to Australia," vol. ii. pp. 25, *et seq.*
‡ T. Worsnop's " The Prehistoric Arts of the Aborigines of Australia."

Those at Chasm Island were discovered by Flinders, January 14, 1803. They were painted with charcoal, and some kind of red paint on white rock as a background, and represented porpoises, turtles, kangaroos, and a human hand. Mr. Westall found, at the same spot, "the representation of a kangaroo, with a file of thirty-two persons following after it. The third person of the band was twice the height of the others, and held in his hand something resembling the waddy of the natives of Port Jackson."* The human figures were nude.

The most notable of the cave-paintings are those found by Captain Grey (the late Sir George Grey),† in March 1838, on the Glenelg River, near the north-west coast of Australia, in long. 125° 9′ E., lat. 15° 57½′ S., and some near the same locality, by Mr. Joseph Bradshaw, in the beginning of 1891, at Prince Regent River, in long. 125° 36′ E., lat. 15° 40′ S., or some thirty-seven miles north-east of Grey's.

There can be but little doubt that all these groups of unique specimens of art—the Depuch Island group is somewhat uncertain—were produced by people of one and the same race, who were foreigners relatively to Australia. One singular characteristic indicates a unity in style of execution, viz., the decoration of the body of certain of the figures with dots. This was a feature of some of the work seen by Grey, Taylor, Cunningham, and Giles respectively.

I shall now restrict my observations to the two most important and wonderful paintings among Grey's discoveries, and the four discovered and sketched by Mr. Bradshaw. Fig. 1 of Grey's was the upper part of a nude (or apparently nude) human form,‡ embracing full face, arms and trunk; the mouth not delineated. or probably worn off the painting. This figure was executed upon the sloping roof of a natural cave, the entrance to which was 5 ft. in height. For the sake of effect the background had been coloured black. The total length was 3 ft. 6¾ in., the

Australasian Association for the Advancement of Science. Brisbane. 1895. The diaper work and medallions figured in the same paper, Plate XII., and occurring at the Ooraminna rockhole on the overland telegraph lines, seem to me either the work of Europeans or done under European influence.

* Flinders' "Voyages to Terra Australia," vol. ii. p. 188.

* Sir George Grey's "Journal of Two Expeditions of Discovery in North-Western and Western Australia," 1837–39.

‡ The artist may have meant to represent this figure clothed with a tight-fitting tunic.

greatest breadth 3 ft. 1½ in., the colouring was in the most vivid
red and white, the eyes being black, a halo of light red was
depicted round the head, its continuity being interrupted by the
neck; triple parallel dotted lines of white crossed the halo from
the head outwards at regular intervals. All round the halo,
rising from its outer rim, there were wavy tongues of flame
done in a darker red. The outline of the halo was dark red,
that of the eyes yellow, that of the nose red. The trunk of the
body, from the level of the armpits down to about the waist,
was marked irregularly all over with red ticks, bearing a strong
resemblance to Sumatran writing.

Grey's Fig. 2 is also of a human form and done on the roof
of another cave. It is clad in a red robe, reaching from the
neck to the ankles, and having tight-fitting sleeves. The total
length of this figure is 10 ft. 6 in.; the face looks right forward; the background of the face is white, the mouth being
indicated by a red streak. No nose appears; the probability is
that the paint has been worn off by the weather. The eyes are
outlined with yellow, which is bordered with a thin red line.
Surrounding the face, there is a broad band of yellow outlined
with red, and outside of this is a broader white band or halo
also outlined with red, and interrupted at the neck. The hands
and feet are coloured dark red. The figure stands nearly in the
military attitude of "attention," the hands, however, being
separated a little from the sides. Immediately over the head
on the outer halo or head-dress are six marks, placed in a
horizontal line at close regular intervals, bearing a general
resemblance to plain Roman letters. To the right of the figure
are three perpendicular rows of small irregular rings, seventeen
in the line next the figure, twenty-four in the middle line, and
twenty-one in the outer line. To the left and close to the
shoulder are two marks which may have been intended for
similar rings. The upper one is like a crescent with convex
side up, the other like a horizontal ellipse, the upper left (to the
observer) quarter wanting. For a view of coloured prints of
these and other paintings seen by Grey, I would refer to his
"North-West and Western Australia," vol. i. p. 201, *et seq.* The
colours employed in both Grey's and Bradshaw's discoveries
were red, blue, yellow, black and white. In Bradshaw's there
was also brown.

I

Various conjectures, some of them rather wild, have been made regarding the origin of this group.   These paintings have been referred to Phœnicians, Spaniards, Portuguese, Japanese and Hindus respectively.   Mr. R. Brough Smyth * thought that, with the exception of Grey's Fig. 1, the authorship of which he considered doubtful, they were the work of natives of Australia, "unassisted by any knowledge gained by intercourse with persons of a different race."   As long ago as 1846, Mr. Hull sought to identify Fig. 1 as Amoun, Cronus or Jupiter.†   He says that one Hindu, who was shown a sketch of it, called it Kons; another called it Koodar or Kadar; and a Victorian black called it Pundyil, a deity of the Victorian natives.   On page 36 Mr. Hull identifies this figure with the Hindu Siva; his conclusion, I believe, is correct, although hardly justified by his premises.   Now, however, we have got fresh light in Mr. Bradshaw's discoveries, and, when viewed in conjunction with them, it is all but certain that this figure is intended to represent one of the Hindu triad, viz., Mahadeva or Siva (the Destroyer Time), who is sometimes portrayed with a halo round the head.

With regard to Grey's Fig. 2, being much struck with the resemblance which the marks on the head-dress bore to the alphabets of Sumatra, I have tried to decipher them, and I believe the result is successful.   By comparing the characters on the painting with the specimens of Sumatran writing, given in V. D. Tuuk's "Les Manuscrits Lampongs," I made out the first four letters to spell D AI B AI; then I found from Marsden's "History of Sumatra" ‡ that the Battas of Sumatra applied the name Daibattah to one of their deities, and that the Cingalese have a cognate name *dewiju ;* the Telingas of India employ the word *daiwunda*, the Baijus of Borneo, *dewattah*, &c.—all to designate a divine being.   I ultimately succeeded in deciphering the whole inscription to read DAIBAITAH.   The following considerations leave, I think, no room for doubt as to the correctness of my interpretation; the authenticated letters are from V. D. Tuuk's "Les Manuscrits Lampongs."   Assuming

* R. Brough Smyth's "The Aborigines of Victoria," vol. i. p. 289.
† "Remarks on the Probable Origin and Antiquity of the Aboriginal Natives of New South Wales."
‡ "History of Sumatra," p. 290.

that Grey copied the painting with perfect accuracy, and that it was in perfect preservation, the characters are :—

$$\mathsf{G} \;\; / \;\; \mathsf{T} \;\; / \;\; \mathsf{L}^{<}$$

My interpretation is *D   AI   B   ÀI   TAH*

$$\mathsf{G} \;\; / \;\; \mathsf{\Lambda} \;\; / \;\; \mathsf{L}$$ ⸴ are *unquestionably* corresponding forms in "Les Manuscrits Lampongs."

Forms of *da* are ⊏, ⊏, ⊏ (and ⊊ (like above) on pp. 56 and 101).

Forms of *ta* are ⊏ ⊏ (⊏ is given by other writers, therefore ⌣ is the persistent part of *ta*).

⋌ is the common form for *ba;* see alphabets, pp. 139–142.

/ is given as *ai* in alphabets of Part V. of MS. A.

In alphabet drawn from Part I. of MS. A (p. 139) there are two forms for 'h,' of which ⸝ placed on the right of another consonant represents *final* 'h,' and, like the vowels with which it seems to be classed, is smaller than the consonants. The character as written will be seen in Part I. of MS. A, and in some cases the lines are almost touching at the angle.

A considerable amount of information is available about this mythical person. The Battaks (or Battas) of Sumatra "believe in the existence of one Supreme Being, whom they name Debati Hasi Asi. Since completing the work of creation they suppose him to have remained perfectly quiescent, having wholly committed the government to his three sons, who do not govern in person, but by Vakeels or proxies."[*] The proxies also get the title of Debata with a modifying word, so that it is the generic name for *deity.* It seems to me that the myth of Daibaitah and his three sons is an accommodation of the Hindu supreme divine essence Narayana with the triad, derived from him or sometimes represented as his modes Vishnu, Brahma, and Siva. The root of Daibaitah and its variants is evidently the Sanskrit Deva (*cf.* Daiva, *fate*), and may be compared with *divus* and *divinity.*

Mr. Bradshaw saw fifty or sixty pictures or scenes. In a paper read before the Royal Geographical Society of Australia, September 10, 1891, when referring to the cave-paintings, he

[*] Coleman's "Mythology of the Hindus," p. 364, *et seq.*

says : " These sketches seemed to be of great age, but over the surface of some of them were drawn in fresher colours smaller and more recent scenes and rude forms of animals." "In one or two places we saw alphabetical characters somewhat similar to those seen by Sir George Grey."

Of Bradshaw's discoveries, Group I. comprises five human figures coloured brown, a snake and kangaroo coloured red, and a legend in characters manifestly of the same type of alphabet as those in Grey's painting. There are also in red two personal ornaments detached : one of these consists of four concentric circles somewhat compressed horizontally, with three discs of like shape, one in the centre and one at each side of the outermost ring ; the other is a band in the shape of half an ellipse, each end terminating in a round disc. There are four spikes projecting from the upper part of this figure, and five others from the rounded end. This is no doubt a sketch of an elaborate and massive earring, as will be shown farther on.

The human figures have long caps on the head.* Three of them have yellow collars, evidently representing gold. One has a girdle with tassels at each side, and armlets at the elbows, from which there are tassels depending. The limbs are poorly executed, both as regards shape and proportion. Total length of scene, from right to left, 12 to 15 feet; greatest height, 6 to 9 feet.

Group II. represents two female figures done in brown. The one is in erect posture, the head turned to one side so as to show the face in profile. The full front of the body is shown, the arms being extended sideways. From the knees downwards has not been sketched. There are armlets at the elbows, and tassel-like ornaments hanging down from the head. The other figure is in an attitude of swimming or perhaps supplication. The side of the body is seen, the hands being extended in front. The figure terminates a little below the waist. Both figures have on long, heavy-looking caps. A crocodile, coloured red, stretches across the picture behind the human figures, its length is about 10 feet, the tail and feet are wanting. The erect female figure is about 5 feet in height.

* What appear to be caps may really be in some of the instances the style of coiffure like that of the natives of Timor Laut, who dress the hair to hang down in a cataract. *Cf.* Forbes "Eastern Archipelago," p. 308.

BRADSHAW

FIG. 1

BRADSHAW

FIG. 2

BRADSHAW

FIG. 3

BRADSHAW

FIG. 4

Group III. contains a bright red figure rudely representing the upper part of a human form. The head is surmounted by nine detached yellow rays. It has three arms or flippers, two red ones where arms would naturally be placed, and extending almost at right angles from the trunk; the third arm is brown, it reaches upwards and outwards from the left side, and at its extremity is a skull coloured brown, with eye sockets and mouth left blank. The body is enlarged and rounded at the lower extremity, which rests on the back of a large serpent, the head of which rises and projects outwards on the left side of the main figure just under the death's head. The serpent's mouth is open, its eyes left blank, the head and neck are coloured yellow, the rest of the body a dark red, the colours meeting in a zigzag line with acute deep angles. In front of the rather amorphous red figure is a human figure without arms. This is of a brown colour, it stands bolt upright and on tiptoe, the feet reach a little lower than the body of the serpent, the head is within the head of the red figure, the latter forming a foil. The brown figure wears a head-dress, has a girdle round the waist, and broad bands or rings on the legs at the knees; from both sides of the head, of the girdle, and of the leg bands, tassel-like ornaments are suspended similar to those already described, giving the appearance of being made of knotted twine, generally with three fringes at the knots, sometimes only one or two ends or fringes. These articles are all the figure wears. From the lower side of the solitary right arm, and from the throat of the serpent, there hang similar tassel ornaments of a dark brown colour. The greatest height of this painting is about 8 feet, the greatest width about 5 feet.

Group IV. has for background what is evidently a huge symbolical painting of a sun-god coloured red. It appears to be placed in a horizontal position; the bullet-shaped head is formed of three concentric circles, with a small disc in the middle. From near the upper part of the head detached red rays extend outwards. The head rests on a short neck, which rests on the middle of the convex side of a crescent-shaped device meant for arms. This consists of two endless bands, one within the other, bent to form a crescent. The concave side of this crescent rest: on the sharply-rounded curve of the outer of three similar bands, one within the other, the six ends forming the termination of the trunk, and completing the symbolical figure.   Drawn perpendicularly across the trunk are portions of four human figures, one complete except feet and arms, another minus feet and having the arms extended upwards in an attitude of supplication, the two others minus arms, neck and feet. Each of the first two has three of the tassel-tipped cords or ribbons hanging outwards from the crown of the head.   All are furnished with belts round the waist, having a tassel at each side.

It seems to me that the most important of these groups are Nos. I. and III.   The characters are of the same type as the Lampong letters, and at once suggest Sumatra as the native home of the artist.   This supposition is confirmed by an inspection of the plates and explanatory letterpress at the end of the "Bataksch-Nederduitsch Woordenbock," by H. N. Van der Tuuk.   One of the figures is an earring worn by women, the resemblance of which to the spiked ornament in the Australian picture is so close as to leave no room for doubt that they both are delineations of the same kind of personal ornament.   And from Van der Tuuk's plate we learn both that the ornament on the Australian picture was not completed and how it would have looked when finished.   In Plate XXII. of the same work there are illustrations giving us a clear idea of what the tassel ornaments in the Australian pictures are meant for.

I have no explanation to offer of the human figures.   I would just draw attention to the fact that the arms of one, if not of two, of the figures are skeleton arms, a pretty sure indication that the picture is symbolical.   The head-dress of the small figure beside the kangaroo is surmounted by what appears to be a head.

The large red figure, with its accessories, is manifestly of religious significance.   It might mean anything or nothing but for the three most striking features—the skull, the serpent, and the rays.   From time immemorial in mythology the serpent has been a token of divinity, ancient statues or paintings of deities were seldom without it.   In Indian mythological paintings Parvati (or Kali or Devi), the consort of Siva, is usually represented as wearing a necklace of skulls, or holding one or more skulls in one or other of her hands, or under both of these circumstances.   One or more serpents are also usually associated with this goddess.   As Parvati she has only two hands; under other aspects the hands are multiplied.

In Moor's "Hindu Pantheon," Plate XXVII., Parvati has a serpent hanging from each ear, one round the neck, and another round the waist. In Plate XXIX., Maha Kali holds a head on the tips of the forefinger and thumb of each of the two upper hands, and in each of the two lower ones she suspends a head by the hair; she also has on a necklace of skulls. This is the most venerated goddess of the Hindus, as being most to be dreaded, and most requiring to be propitiated.

Plate XXIX. gives an Avatara of Siva, seated on the folds of a serpent, whose head surmounted the god's head. This figure has four hands, in one of which she holds a head by the hair. Another mark of Siva is a halo round the head. In Coleman's "Mythology of the Hindus," p. 91, Parvati is represented under the form of Kali, the consort of Siva, in his destroying character of Time. In Plate XIX. she is shown as a personification of Eternity trampling on the body of Siva, her consort, Time; in one hand she is holding a human head. Hindu pictures in which the god is represented seated cross-legged, with his consort on his lap and his arms around her, are frequently to be seen.

These references should, I think, be sufficient to identify this picture as a combined representation of Siva and Kali.

A conjunct view of these paintings leaves no doubt as to the nationality of their authors, and the significance of the best of them is tolerably clear. It is obvious that there has been an attempt to present pictorial fragments of Hindu mythology in the confused form which has been developed by naturalisation in Sumatra. The attributes of both Siva and Kali his consort are allegorically expressed, whatever the names may have been by which these deities were known to the artists. Daibaitah, with his three sons and their proxies, may be a version of the Hindu triad which has been freshly elaborated, perhaps unconsciously, by the Sumatran mind. One is naturally curious to discover what the three rows of rings on the right of Daibaitah and the two marks on his left may symbolise. In these, also, there is an imitation of Hindu sacred allegorical art.

In Moor's "Hindu Pantheon," on Plate XL., there is a figure of Devi, at the side of which there are two perpendicular rows of oblong marks, some oval, some rectangular, five in each row.

On the same plate Bhavani (or Devi) is represented with two perpendicular rows of oval marks, three in each row. On Plate LXI. two different representations of Devi have, round the border, the one a string of imperfect circles, the other a string composed partly of circles, partly of squares. A third picture of her on the same plate has a border of circles at the bottom, and near the head are a crescent on the right and a circle on the left, manifestly symbolising the moon and the sun. Other plates of Devi have rows of rings at the bottom, with a crescent and a circle near the head. There is doubtless as close a relation between the circles on the Australian pictures and those on the Indian ones as between the names Daibaitah and Devi. They indicate attributes of the particular deity.

Dr. Adam Clarke says that the o o o among the Hindus is a mystic symbol of the deity signifying silence, which seems scarcely an adequate explanation. Perhaps the inference that the two marks standing apart near the left shoulder of Daibaitah may symbolise the sun and moon is rather bold and unwarranted, but from comparison with the Hindu pictures one cannot help a surmise that this may be the case.

The artists of these extensive works must have spent an immense amount of time and mental and physical energy in their execution, the first impulse being probably imparted by religious feeling. One cannot but believe that there was a number of sacred men among the immigrants from Sumatra, and that some at least of these caves, upon the decoration of which skill and labour were so lavishly spent, were shrines where worship was offered. Just under the roof-tree in Sumatran temples (?) (*Sopo*), there is a carving of a human head called 'buwaja-buwaja,' *i.e.*, the figure of a crocodile, because in earlier times, and even still in primitive places, instead of a man's head the figure of a crocodile is placed in that position *—this is interesting as suggesting a sacred meaning attaching even to Bradshaw's Group II.

Whatever influence these religious foreigners may have exerted in the neighbourhood of the Glenelg and Prince

* " Bataksch-Nederduitsch Woordenbock " (H. N. Van der Tuuk). Letter-press at end of book explanatory of Plate II.

Regent Rivers, it seems to have all but faded away. Mr. William Froggatt, of Sydney, New South Wales, visited the Glenelg River in 1887-8. The aborigines could give no satisfactory account of the paintings, but said they were pictures of the "Nauries," black evil spirits, the agents of all ill and of whom they were afraid. This ignorance as to the origin of the pictures goes to show that they must have been done at least a hundred years ago. Mr. Froggatt states that the natives in the locality wear "tails" on the forehead to keep away the flies and waistbands made of opossum fur or human hair, which adornment may correspond to some shown on the figures. As regards the "Nauries," Mr. Joseph Bradshaw informs me that "the only religious ceremony practised by the Yuons (in Kimberley, north-west Australia) is an occasional corroboree in honour of Nari, of whom they cannot or will not give much information, but ascribe to him (or it) the creation of all things long ago." The name Nauri may prove a means of obtaining further light upon the relation between the Sumatrans and the Australians of the north-west coast.

The rite of circumcision was probably introduced to Australia by Sumatran natives, a view which is confirmed by local distribution of this practice. The making of hand-prints upon rocks in red mostly, but sometimes in black, which may be considered universal in Australia, is probably a practice derived from the same source, for Captain Grey (now Sir George Grey) saw a hand and arm done in black, and Flinders' party saw a hand painted presumably in red. In India the print of a hand is said to be emblematic of taking an oath. Mr. Curr has seen the blacks making such impressions for pastime, and he is of opinion that others which have been observed may be also modern, and of no special significance. From the occurrence of these "red hands" in places very far apart and from the peculiar position and arrangement of groups of them,* I cannot help concluding that they are in the first instance sacred symbols, however frivolously they may have been imitated by blacks who had lost the meaning of them.

It is not too much to expect that careful investigation may demonstrate the religious beliefs and sacred or mysterious rites

* Dr. Carroll's paper, "The Carved and Painted Rocks of Australia." *Centennial Magazine*, October 1888.

of the Australian aborigines to have been largely inspired and shaped by the settlement of people from the island of Sumatra deeply imbued with religious feeling.

It is only reasonable to believe that the higher class of paintings so skilfully executed and so mysterious and varied in subject have influenced the aboriginal mind towards some degree of imitation. At any rate, caves and rock-shelters, in other quarters remote from the superior work, are found to be covered with much ruder sketches of men, animals, weapons, and symbols. Whatever secret meaning these may possess has not yet been discovered. Various parts of New South Wales are rich in such memorials, specially the neighbourhood of Singleton, in the county of Northumberland. My friend Mr. R. H. Mathews, Mr. W. J. Enright, and others, are enthusiastically engaged in copying these remains. Very few specimens have as yet been found in Victoria. The writer copied one group depicted on the face of a huge rock-shelter, on the Billaminah Creek, in the Victoria Range, near Glenisla, Victoria. The width of the smooth face of the rock is about 50 ft., the height over 60 ft. The greater part of this southern aspect up to a height of from 6 ft. to 8 ft. 5 in. is covered with figures. The sketching is done in red, either dry with a very fine-grained red sandstone, or with the same material powdered, or a red earth mixed with opossum fat. I have given a full account of the painting in the "Proceedings of the Royal Society of Victoria, 1896," in which also the principal drawings are figured. The figures are not so large as some of those on the rocks in New South Wales. A sample is shown here.

In the neighbourhood of Sydney there are numerous carvings of animals and other objects upon the sandstone rock. Men, fishes, boomerangs, spears, hatchets, are all delineated, generally not in single representations but in groups, and manifestly with the aim of conveying some kind of knowledge. Dr. Carroll, referred to above, undertakes to explain these, but his interpretation is clearly mere conjecture, and has little to recommend it beyond possibility. When, for instance, he distinguishes between ancient and modern carvings by the fact that one set is overgrown with mosses, and the other not, he is plainly quite astray. Whether a stone be bare or clothed with moss or lichens, after the lapse of, say, fifty years from the time

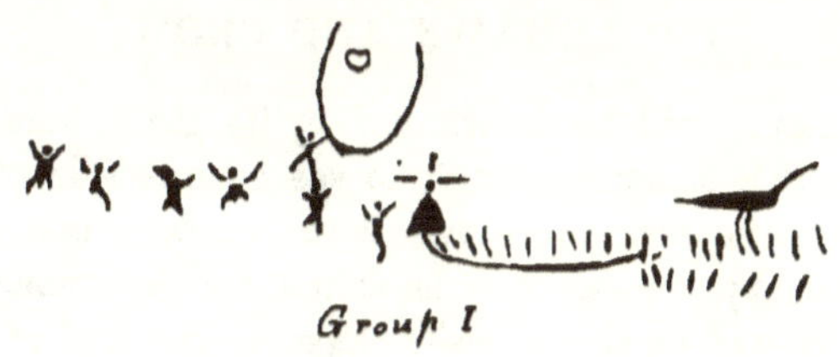

Group I

Group II

Group III

ABORIGINAL ROCK PAINTINGS IN THE PARISH OF BILLAMINAH
COUNTY OF DUNDAS, VICTORIA

of its exposure will be determined by its grain, hardness and
position. It is common enough to see on one side of a road-
cutting basaltic rock overgrown with lichens, while the same
rock on the opposite side may be naked, and a similar contrast
may be observed on the opposite slopes of the roof of a house.
That these rock-carvings were symbolical is almost beyond
question, and they have parallels in workmanship, although not
in subjects, in carvings that occur upon rocks in the South Sea
Islands. Sir George Grey also speaks of a head carved by scooping
the rock, seen by him near the caves on the Glenelg River.

The Australian muse is cultivated enthusiastically but un-
progressively. The native in this respect, as in al respects,
is conservative to the backbone, so that we have no reason to
suppose that his music and song of to-day are any advance upon
what they were three thousand years ago. In some cases the
words seem designed to run in rhymes, but a decided rhythm
recurring in lines of regular length and invariably chanted,
never recited, is the essential character of Australian poetry.
Almost every blackfellow is a "maker" of lyric verse, and whiles
away the hour with his own compositions about any subject
which lies closest to his heart, but the man who has the talent
to compose a dramatic corroboree is a person of no small
consequence.

The rendering of a corroboree is an occasion of the most
intense enjoyment. The males are usually the sole performers,
the women sitting in front by the fires and beating time by
striking two sticks together or clapping with their hands upon
stretched opossum skin or on the hollow of the thighs formed
by the sitting posture. The dancers are smeared with red or
yellow clay, or with pipeclay, in patterns that give them a
frightful and sometimes a ghastly appearance. Occasionally
the limbs are decked with light sprays. A customary movement is a shaking of the legs and a wriggling of the body.
There is a tramping time to the music and a scarcely recognisable
representation of some action. Some corroborees are lewd in
the extreme, and it is generally understood that at such times
sexual restrictions are shamefully, or from the native point of
view shamelessly, relaxed. Popular corroborees are transmitted
from tribe to tribe and sung where not a word of them is understood. The melody moves at times very *andante*, and the very

same sounds, after a signal given, will be sung in the liveliest manner. The modulation is exceeding easy and gradual, the music rising or falling by semitones, save when, after a gradual descent, there is a sudden vocal leap of an octave upwards. The close of a piece will be indicated by three great yells, which do duty for the crashing music indispensable to finish off most of the compositions in the *répertoire* of civilisation.

—

CHAPTER XI

SORCERY, SUPERSTITIONS, RELIGION

Sorcery, superstitions, religion — The bane of sorcery — Native magicians or doctors—Their professed powers—Native phlebotomy— The rainbow—Spells—Names of deceased persons—Sacred pebbles— Ghosts—Ancient heroes—Deities.

THE greatest bane of aboriginal life, as of all savage life, is sorcery. People reared in absolute ignorance of its bloody tyranny are unable to understand why the old Mosaic enactments should be so severe against its practice, but the necessity for such severity lies in the diabolical character of the thing itself and the proneness of the human mind to submit to its thraldom. It may be truthfully affirmed that there was not a solitary native who did not believe as firmly in the power of sorcery as in his own existence ; and while anybody could practise it to a limited extent, there were in every community a few men who excelled in pretension to skill in the art.

The titles of these magicians varied with the community,* but by unanimous consent the whites have called them "doctors," and they correspond to the medicine-men and rain-makers of other barbarous nations. The power of the doctor is only circumscribed by the range of his fancy. He communes with spirits, takes aërial flights at pleasure, kills or cures, is invulnerable and invisible at will, and controls the elements. I remember a little black boy, when angry, threatening me with getting his father to set the thunder and lightning a-going. The same boy told me seriously that on the occasion of a raid being made upon the blacks' camp by the native police, one of his fathers—a doctor—lifted him and pitched him a mile or two into the scrub, and vanishing underground himself, reappeared

---

* *Koradji* was the name applied in the neighbourhood of Sydney, and it still holds the ground among Europeans.

at the spot where the boy alighted. The doctor has great
healing skill. A common exercise of this is to extract some
object from the seat of pain by sucking. The object may be a
piece of glass, or a plug of tobacco, or a half-brick.

The *manngur* (plu.), *i.e.*, the doctors of the Kabi tribe,
followed a practice of fictitious bleeding, known also in other
tribes, for the relief of pain. The *manngur* was provided with a
long cord made of fur and a vessel containing some water.
One end of the cord was fastened round the body of the sick
person immediately over the seat of pain, and the other end
was placed in the water. Seating himself between the patient
and the water-vessel, the *manngur* held the cord about the middle
with both hands and rubbed it backward and forward across his
own gums, causing them to bleed. As the saliva and blood
accumulated in the mouth it was expectorated into the vessel.
The process was carried on in a slow deliberate fashion until the
water in the vessel became quite discoloured. The blood in the
frothing liquor was supposed to be drawn from the patient, who,
at the close of the operation, had to drink the contents of the
vessel.

I do not know whether there is any uniformity of belief as
to what confers the special gift of sorcery, but the opinion of
the Kabi community (Queensland) was distinct enough. The
doctor might be, as one might say, of two degrees—a ' kundĭr
bŏnggan,' a sort of M.B., and a 'manngŭr,' a thorough M.D.
A man's power in the occult art would appear to be proportioned
to his vitality, and the degree of vitality which he possessed
depended upon the number of sacred pebbles and the quantity
of *yurru* (rope) which he carried within him. One kind of
sacred pebble was named ' kundĭr,' and the man who had an
abundance of them was called ' kundĭr bŏuggan ' (*pebbles many*),
and was a doctor of the lower degree. The ' manngŭr' was a
step in advance. He had been a party to a barter with
' dhakkan,' *the rainbow*, and the latter had given him so much
rope for a number of pebbles, which he had taken from the man
in exchange. This transaction took place while the black was
in a deep sleep. He would be lying on the brink of a water-
hole—the rainbow's abode. The rainbow would drag him under,
effect the exchange, and deposit the man, now a ' manngŭr
manngŭr,' on the bank again. The doctor carried his sacred

apparatus in a small bag, which none but himself might venture to touch, for fear of sudden death. He could hang the bag up anywhere in full view, perfectly secure from interference; and he slung it on his shoulder when on the move. Its contents would be a few pebbles, bits of glass, bones, hair, cord made of fur, and perhaps excreta of his foes. Certainly not a very formidable artillery, but for him enough to kill at any distance. In fact, sickness and death were usually attributed to the practice of sorcery. A blackfellow gets a stitch in his side, and immediately he believes that an enemy has cast a pebble at him from behind a tree. The reasoning process is simple. The law of causation requires a cause for everything, and as a man would not get ill of himself, an enemy must be at the bottom of any hurt which he sustains. There was an interminable process of mutual revenge going on between neighbouring tribes, and the blow of the club would respond to the challenge which had come in the form of sickness from sorcery.

There are, or were, numerous superstitions of the nature of religious belief, inasmuch as they acknowledge invisible supernatural powers and beings. The blackfellow lived and moved and had his being in superstition. Unseen instruments and agents were continually at work. Disease would result from violation of rules, as, for instance, from eating prohibited food. To obtain possession of a person's hair or ordure, was to ensure his death. He declined as these decayed. It was dangerous to pass under a leaning tree or fence. The reason alleged for caution in this respect was that a woman might have been on the tree or fence, and that blood from her might have fallen upon it. This would seem to point to former regulations regarding pollution. But it may be the Australian form of the *mana* superstition, said * to be a sure mark of the Polynesian race.

Akin to this dread of passing under an elevated object, and due no doubt to the same cause, is the fear of another person's stepping over one's body. Both these superstitions indicate the belief that a baneful influence of some kind is liable to fall from a person, and this influence was probably supposed to be due to some emanation like the *mana* of the Fijians.

The objection to pronounce the names of dead people has

* By Rev. Dr. L. Fison, M.A., *Centennial Magazine*, February 1889, p. 457.

been noticed by most writers on the Australian race. The aversion would seem to be the result of a kind of *realism* among the natives, whereby a person's name became through confusion of thought the same as himself.

The veneration of pebbles has already been noticed. It has been remarked that the blacks were exceedingly loth to permit white men to see their sacred objects, and they were also concealed from their own countrywomen. There were local preferences for certain kinds of pebbles, but in general they appear to have been simply smooth rounded quartz stones.

The Rev. J. G. Paton secured a small piece of wood painted red at one end which he says is similar to one kind of idol worshipped by the New Hebrideans. Mr. Taplin describes * a practice of sorcery, called 'ngadhungi,' followed among the Narrinyeri, which bears upon the significance of the piece of stick coloured red at one end. A bone forming the remains of a repast of some native is secured and scraped. "A small lump is made by mixing a little fish-oil and *red ochre* into a paste and enclosing in it the eye of a Murray cod, and a small piece of the flesh of a dead human body. This lump is *stuck on the top of the bone* and a covering tied over it, and it is put in the bosom of a corpse that it may derive deadly potency by contact with corruption; after it has remained there for some time it is considered fit for use. Should circumstances arise to excite the resentment of the disease-maker towards the person who ate the flesh of the animal from which the bone was taken, he immediately sticks the bone in the ground near the fire, so that the lump may melt away gradually. The entire melting and dropping off of the lump is supposed to cause death." Could human ingenuity be exercised in a manner more sickening, horrifying, and repulsive? A similar demand for the remains of food or other refuse of what a person has used is a trait of South Sea Island superstition. Although there is great dissimilarity in language between the Polynesians and Australians, such common traits as a community in objects of worship bespeak a close connection at some time. History proves how easily a form of worship may be superposed upon existing forms, whereas it usually requires violent causes to change language by the substitution of one tongue for another. It may therefore be

* "Native Tribes of South Australia," p. 24.

K

the case that such resemblances in superstitions are due to independent similar transitory causes, or say, to the drifting of a few Kanaka canoes to Australian shores, although from the fact that stones were objects of veneration among the Tasmanians the inference would be that this at least was a superstition common to all primitive Papuans.

The Australians have what may be termed an apprehension of ghosts rather than a belief in them, the relations of the living with the spirits being more or less intimate in different tribes. In the tribe with which I was best acquainted, while the blacks had a term for ghost and believed that there were departed spirits who were sometimes to be seen among the foliage, individual men would tell you upon inquiry that they believed that death was the last of them. In other words, a man's personality died with his body and was not continued in his ghost. A ghost was called a 'shadow,' and the conception of its existence was shadowy like itself. A general feature of Australian mythology is the peopling of deep waterholes with indescribable spirits. The Kabi tribe deified the rainbow, a superstition apparently confined to this people. He lived in unfathomable waterholes on the mountains, and when visible was in the act of passing from one haunt to another. He was accredited with exchanging children after the fashion of the European fays. He was also a great bestower of vitality, which he imparted in the form of 'yūrru'—*i.e.*, *rope* (what this rope was I do not know), in the manner explained above.

Many tribes revered the names of ancient heroes or demigods, who were credited with certain wonderful exploits, and who generally became metamorphosed into stars. The conception of a supreme being oscillated between a hero and a deity. Some tribes recognised both a supreme good spirit and a powerful, dreaded, evil spirit, creation being ascribed to the former. At the initiation ceremonies of the Darkinung tribes of New South Wales * two figures are made upon the ground by heaping up earth. They are represented as like human beings lying flat on the back. A quartz crystal called *ngooyar* is placed upon the forehead of Dhurramoolun, the good spirit, and a koolaman (wooden vessel) containing blood just let from the arms of some men is placed upon the breast of "Ghindaring,

* R. H. Mathews, "Proc. Roy. Soc. of Victoria," vol. x. part i. pp. 2-3.

a malevolent being," whose body is said to be red and to resemble burning coals. I was once of opinion that notions about a divinity had been derived from the whites and transmitted amongst the blacks hither and thither, but I am now convinced that this belief was here before European occupation. Although not entertained by every tribe, it is nevertheless held by one tribe or another in the south-east quarter of the continent, from the coast to the centre, and we are justified in concluding that it extends beyond the area where it is positively known to exist.

By those who have been eager to establish the theory that there are atheistic races of men the Australians have been cited as an example, another instance of the unreliability and invalidity of a deduction from negative evidence. Among the Kamilroi and allied tribes to the north of New South Wales the character of a beneficent deity, known as Baiame, has been well elaborated. The name, according to the Rev. W. Ridley, is derived from ' baia,' *to make or build by chopping.* Baiame is the creator and preserver. The Wiradhuri regarded him under a slightly altered name, Baiamai, as eternal, omnipotent, and good. A supreme deity was known by the name of Anambu or Minumbu by the Pikumbul tribe; at Illawarra he was called Miriru; on the Murray Nourelle, in Victoria, he was generally known as Bundjil or Pundyil, and also as Gnowdenont; the Narrinyeri, as we have seen, called him Nurundere and sometimes Martummere, and by the Diyeri he was known as Mooramoora.

Dr. Lang [*] observes, "There are certain traditions among the aborigines that appear to me to have somewhat of an Asiatic character and aspect. Buddai, or, as it is pronounced by the aborigines towards the mountains in the Moreton Bay district, Budjah (*quasi* Buddah), they regard as a common ancestor of their race, and describe as an old man of great stature, who has been asleep for ages." The question may be reasonably asked is this Buddai not as likely to refer to Daibaitah of the northwest as to Budha? In New Guinea, according to Marsden, the same deity is known as ' Wat,' the first and third syllables of the name being lopped off. And further, may it not be possible that Baiame, of New South Wales, and Pundyil, of Victoria, refer to the same supernatural being? Baiame, indeed, may be a local

* " Queensland, Australia," p. 379.

equivalent of Barma, another Sumatran deity. The blackfellow Yangalla already mentioned recognised Daibaitah as Pundyil; the fancied resemblance may have been due to an impression that both were supernatural beings, but, on the other hand, the names may be etymologically related, and if so, a unity is given to the native belief in a divinity.

The myths regarding the creation are numerous, and there are some which refer to a flood, but there is no common fixed account of either event, and both classes of myths may be quite modern, the one being an attempt to explain the world's origin, and so far a reflection of the workings of the native mind, the other a recollection of an unprecedented local downpour of rain and consequent inundation. I confess to having failed to obtain in the south of Queensland any myth about the creation or the flood; the nearest approach to an account of the former was the personal conjecture which a blackfellow made regarding the origin of his race, which was that he thought they had sprung up like the trees—uncommonly like Topsy's "I specs I grow'd." The Arunta tribe in central Australia have an intensely interesting myth about the 'Alcheringa,'* the earliest period to which their traditions refer. "At the very beginning of this there were no true human beings such as now exist but only 'Inapertwa,' that is, almost shapeless beings in which just the vague outlines of the different limbs and parts of the body could be detected. Two spirit beings who lived far away in the western sky and who were called 'Ungambĭkŭlla,' a word which signifies 'made out of nothing,' or 'self-existing,' came down to earth and transformed the Inapertwa creatures into men and women." The men and women of the Alcheringa are also said to be "direct descendants or transformations of animals" whose names they respectively bear.

* "Notes on certain of the Initiation Ceremonies of the Arunta Tribe, Central Australia," by Prof. W. B. Spencer and Mr. F. J. Gillen, "Proc. Roy. Soc. Victoria," p. 146, *et seq.*

CHAPTER XII

AUSTRALIAN LANGUAGES

THERE is need for extreme caution in making sweeping general statements regarding the languages of Australia, because they are liable to be invalidated at any time by the discovery of particular contradictory instances. The judgment of the very highest authorities is subject to be shaken by exceptions. For instance, in the " Reise der Fregatte Novara," p. 244, Dr. F. Müller says that the aspirates 'h' and 'v' are wanting, whereas there are undoubted, though rare, cases of their occurrence. On the same page he says that these languages possess only post-positions, and that they have no distinction for gender; whereas there are instances of prefixes and infixes as well as post-positions, and some dialects have phonic marks to distinguish sex. Again, Dr. Codrington,* referring to the numerals in use on the islands of the Pacific, New Guinea included, says, " Any of these island numerals will be looked for in vain on the continent of Asia, Africa, or Australia." Now I am able to trace numerals from the most southerly part of Australia northward into New Guinea and even to Woodlark Island (*see* table where Australian and New Guinea numerals are compared). The term for *two*, ' luadi ' (not in the table), used by the fierce Kalkadoon tribe in north-west Queensland, is certainly an island word. When, therefore, any character is affirmed to be universally

* " The Melanesian Languages," p. 243.

true of Australian languages, the statement must imply the
reservation that there may be exceptions, until at least all the
various dialects have been reduced to writing and brought
under survey.

Dr. Bleek has classified the Australian languages in the
following manner:—

I. Northern Division.

II. Southern Division   {   1. Western.   2. Middle.   3. Eastern.

III. Tasmanian.

No exception can be taken to giving Tasmanian dialects a
place by themselves, but the other part of the classification is
too loose and arbitrary. It should be borne in mind that at
least one-third of Australia is a *terra incognita* to the philologist,
the unknown part embracing nearly all the western half of the
continent excepting a strip along the coast. Very important
ethnological revelations may be awaiting disclosure there.

As a geographical classification based upon present knowledge, I would suggest the following, which corresponds, as
regards the larger classes, to the arrangement in the Comparative Table at the end :—

I. Tasmanian. Subdivisions : (1) Eastern; (2) Central; (3)
Western.

II. Victorian Region, embracing part of Riverina and Murray
Basin in N. S. Wales, also south-east corner of South
Australia. Subdivisions : (1) Eastern (Gippsland) ; (2)
Western.

III. N. S. Wales and South, Centre, and East of Queensland.
Subdivisions : (1) Eastern (Coast) ; (2) Western (Inland,
west of Dividing Range).

IV. W. Australia and West Central. Subdivisions : (1) Northern ;
(2) Southern.

V. South of S. Australia and East Central, including West of
N. S. Wales and North-West of Queensland.

VI. North Coast and Central Australia, including C. York
Peninsula and North-West Coast. Subdivisions : (1) Coast;
(2) Central.

Notable diversities in words and structure are due in the
main to dissimilarity of original elements, while the shading of
dialects into one another must be ascribed partly to the influence
of exogamy, partly to a very gradual change of old elements,
partly to the introduction of fresh words from the north. Almost
everywhere throughout the continent original elements are

observable protruding through the more recent, like primary rocks through all later formations.

If it be asked, what view of Australian settlement does a study of language lead to? the reply must be that a general survey of the languages favours the conclusion that there was at first a comparative homogeneity of speech in Australia and Tasmania of simple structure, as exhibited in Tasmanian and Western Australian dialects, and that subsequently there poured in from the north-east streams of population with a speech more elaborate in construction. Parallel strips of homogeneous people still testify to this migration from the northward, and the general brokenness of language along the east coast betokens disturbing and overlying influences from the landward streams, and in some cases most certainly through settlements of people who have come not overland, but by sea.

The fundamental principle of word structure is agglutination. There is therefore a general well-marked relationship with the members of the Turanian branch of human speech, with at least one contrast, viz., the absence of the law of Vocalic Harmony. The usual form of modification is by post-positions, but this mode is by no means invariable; it is supplemented in many dialects by preformatives, and sometimes by included particles. Genuine internal vowel inflection is not observed. Certain languages lean almost as much to external inflection as to agglutination and tend to analytic structure. Others are fairly perfect specimens of agglutination. So far as I know, the simplest and most analytic language is that current in West Australia; the most complex, if it have not a rival on the Daly River, is at Lake Macquarie, at the other extreme of the continent; and between the east and west extremes the languages increase in complexity and fineness of elaboration from west to east. This does not involve the inference that a process of modification proceeded latitudinally along the continent, because, as a matter of fact, the language of the south-west crossed directly from the north-east. The speech of Western Australia might be taken as the extreme of simplicity, a dialect of New South Wales or Southern Queensland as the extreme of elaboration, while the language of the Diyeri, lying about half-way between, is simple in structure but richly compounding.

In nouns and adjectives there is a conspicuous abundance of dissyllabic words, as in the South Sea Island languages. In many cases I am convinced that these are a combination of two roots, the original sense of one syllable being lost or the sense of both transmuted. For instance, take the word 'wulwi,' *smoke* in the Kabi dialect; neither of the syllables separate has any meaning in Kabi, but in other parts 'wolla' is *rain*, and 'wi' is *fire*, so that 'wulwi' meant originally *rain of fire*. Decomposition after this manner would no doubt throw much light upon the primitive speech.

PHONIC SYSTEM.

The phonic system embraces all the pure vowels; the modifications expressed in German by ä, ö, and ü being also used by certain European writers. All possible diphthongal combinations are employed in one dialect or another. The consonants found invariably are k, t, p, ng, ñ, n, m, y, w, r, l. Owing to a common imperfect enunciation of the mutes, some are doubtful whether g, d, and b can be credited to Australian languages. The most certain proof, to my mind, that these sonants should be included in a complete summary is a remarkable unanimity in foreign ears recognising them in certain words, as for instance in 'bulla,' *two*, 'barang,' a class name. Besides the foregoing, there are the aspirates 'dh,' 'th' (as in English father), and 'v'; there is a cerebral 'r,' which I shall mark 'rr,' a conjunction of 'dy' and of 'ty' approaching so nearly to English 'j' and palatal 'ch' as to be expressed sometimes by these. The aspirate 'h' occurs, but is extremely rare, and the rushing sonant 'gh' is written in some Victorian dialects, as also in Tasmanian. Sibilants are occasionally given, but their actual occurrence in pure form requires confirmation.

An aversion to 'r' and 'l' as initial letters is very common. In several New South Wales and Queensland dialects these letters never begin a word. Introductory vowels are also generally avoided; if occurring in considerable number in any dialect the peculiarity becomes a differential feature. Any of the consonants employed in a dialect, other than the liquids 'l' and 'r, may commence a word. The nasal 'ng' and a consonant followed by a furtive 'y' like 'ly,' 'ty' (mouillé consonants), are

common initial letters. Initial combinations in words or syllables, such as 'kr,' 'gr,' 'dr,' are generally disliked; and where
frequent, as, for instance, in Victoria, are primitive Papuan
marks. In the dialects of New South Wales, southern Queensland, and central South Australia, such forms are exceedingly
rare, and even where they appear to exist there is a disposition
to interpose a short vowel. I am confident, from comparisons
I have made, that originally a vowel separated those consonants
that are now contiguous. In the smoother languages of the
east the possible terminal letters are usually limited to the
liquids, 'ng,' 'ndh' (a dentated 'n'), and vowels. In Victoria
and West Australia words may end in any consonants and
vowels, and such terminal combinations as 'rk,' 'rn,' 'rt,' are
not objectionable; another evidence of the closer alliance
between the languages of these places. Another rough mark
of relationship in the Victorian and West Australian languages
is the admission of closed syllables in any part of a word, an
exception to the general rule, which is a preference for open
syllables, unless either a liquid be the closing letter or the final
letter of one syllable be also the initial letter of the succeeding,
an exceedingly common character, in which case both letters are
distinctly enunciated, *e.g.*, 'kokka,' 'kakkal.' The accent
usually falls upon the first syllable, which is also the main
radical. A circumstance which has greatly hindered our
acquisition of a perfect knowledge of Australian languages is,
that unless the inquirer be a philologist, we bring in to the
study our European ideas of language, and endeavour to force
aboriginal forms into European grammatical moulds, a practice
which should be at once discarded, as it cramps all inquiry.
It does great injustice to the native languages, on the one hand
making them appear extremely rudimentary, and on the other
leaving many verbal forms unexplained.

POINTS OF CONTACT BETWEEN AUSTRALIAN AND NEW HEBRIDEAN LANGUAGES.

Hitherto little or nothing has been done in the way of
showing to what extent the Australian and New Hebridean
languages are related. The agreement in some grammatical
peculiarities has been noted, but the existence of important

TABLE OF ANALOGIES.

Note.—New Heb. often adds -na, *his*. The Malay equivalents are
introduced for comparison.

ENGLISH	MALAY	TASMANIAN	AUSTRALIAN	NEW HEBRIDEAN (CHIEFLY TANNESE)
father . .	bapa	—	aba, baba, yabu, bapo; mam, mama, ama	abab; babu(voc.); mama (voc.)
to be . .	—	—	nyenni (Kabi, Q.)	ani, eni
man . .	—	—	dhan (Kabi, Q.)	ata, ta
to come . .	—	—	ba(imper Kabi,Q.)	ba (*to come from*)
belly . .	—	ploner, ploang-ner (*stomach*)	polloin, belanyin, beleni (*stomach*, Vic. Reg.)	bala, bele, balau
to kindle .	—	parik	baraio (Kabi, Q.)	buria, bauria; bara (*to be burned*)
big, fat .	—	paroi, proi	parok, parronk (Vic.)	barna
head . .	—	poyta	bua, poko, bo, baukine (Vic.)	bau-na
this . . .	—	(1) nara; prob. also affix -na	(2) karina (Kabi, Q.)	(1) nga; (2) ka
nose . . .	—	—	nogro, noor, no-gooroo (N.S.W.)	ngore-na
small . .	kachil, kete	keeta, kaeete	kutchulka and variants (Dar-ling R.)	(1) kita; (2) kiki
face (cf. *nose*)	—	—	ngoo (Kabi, Q.); ko (*nose*)	ko
dog . . .	—	—	kulli (Q., N.S.W.); kadli (S.A.); kal, kalli (Vic.)	kori, kuri ; other dial., kuli
skin . . .	kulit	—	ula, yuli, yulin (N.S.W.)	kuli-na, uili-na
tomahawk .	—	—	kodja (W.A.); kootya (Wilson R.); koocha (Up.Cape R.,Q.)	kut-ia (*to cut*)
wind . .	—	ıawlin-na, rallin-ganunne	willangi (Vic.)	langi, c. art. na-langi
woman . .	—	loa, lowa	laioo, laioor, laurk	lai, lei, le, li
light . .	—	loina (*sun*)	arlunya, allunga (*sun*, Central Australia)	lina
to be black	—	loarra (*charcoal*)	lourn (*charcoal*, Gippsland, Vic.)	loa
man . .	—	—	mari, murri (Q. and N.S.W.); maar (Vic.); mean, main (N.S.W.)	ma'ani, mani, mare(all mean-ing *male*)
alive, to live, animal	—	—	moron, murree, murru (*full of life*); murang (*animal*) (Kabi, Q.)	mauri (*to live*); maurien (*life*)

TABLE OF ANALOGIES—*continued.*

English	Malay	Tasmanian	Australian	New Hebridean (chiefly Tannese)
to do ..	—	—	-mathi, -maraio (verbal terminations, Kabi, Q.)	mer ia
eye ...	mata	moygta	mi, mil, mir, miki, mityi	mita, meta
tongue ..	—	mena, mene	mooni (S. of Gulf of Carp.)	mina
water ..	—	moga, moka, mookaria	ngok, nokko, mookooa; mukkera, mookorar (*rain*)	mok (*water from the eye*); ura (*dew*)
this ...	—	-na (prob. art.); nara (*this that, he, they*)	ngerma (*he, she*); ngarma (*they*)	na (prefixed art. postf. *this*); naga, ra (*this*)
they ...	—	nara	(1) ngerma, ngarma (Kamilroi, N.S.W.); (2) tana, thana	(1) mara; (2) ta (nom. and verb. suff. 3 plu.)
excrement .	tai	tinmena, tiannah	gunda, goodna, dagga	tai
man (vir) .	—	—	giwir, kivar, kippa (after initiation)	takuwer (fr. ta *man*; kuwer *big, strong*)
earth ..	tanam	gunta, coantana	taon (N.S.W.); dha (Kabi, Q.)	tan, tano
leg ...	—	tula	tirra, dhirang	tere (*calf*)
feet ...	—	tyentiah, teeantibe (*to trample*)	dinna, dhinnang, tinna, &c.	tua na
where ..	—	—	wendho	uan ue (to, *to rest, dwell, be*)
axe ...	—	—	wakko, wokkaka, wakkan	noki
mallee hen .	—	—	lowa (Vic.)	lawaig (*bird like a hen*)
louse ..	kutu	—	kutta (Daiyeri, S.A.)	kutu
interrogative sign of plural	— / —	wanarana / —	wa- or we-ngilpung (*you two*); ngilpulla (*you*); ngalpa, ngalpu (*we three*, Saibai, Torres Strait)	na, ue / ilpu (*the, these*)

glossarial analogies seems not to have been suspected. For this
failure or omission the absence of published data is no doubt
mainly responsible. Dr. D. Macdonald's " Etymological Dictionary of the Language of Efate " throws much new light upon
the new Hebridean languages. From a perusal of this work I

have been able to detect several analogies with Australian speech, a few of which are of great interest, and should prove helpful to further research. An older work, a Dictionary of Aneityumese, by the Rev. John Inglis, also adds to our knowledge in the same direction.

Several of the above analogies demand special notice, not only as tending to show a radical connection between Australian and New Hebridean dialects, but also as suggesting a closer kinship than has been supposed to exist between some of the racial constituents in the two regions.

(1) The terms for *father* are practically identical. Dr. D. Macdonald explains Efatese 'mama' as the vocative of 'abab.' If he is right, we thus reach the important solution of the same form ('mama') which in Victoria and West Australia appears as a word distinct from its so-called nominative occurring elsewhere in Australia.

(2) The terms for *belly, head, black, woman, light, water, they, excrement, earth, feet* unite the Australian, Tasmanian, and New Hebridean languages. Of these the first five are peculiar to Tasmania, the Victorian Region (Australia), and the New Hebrides.

(3) The terms for *small, eye, excrement, earth* seem to be common to the four compared languages.

(4) The terms for *father, skin* are the same in Malay, Australian, and New Hebridean.

(5) Besides explaining 'mama' as vocative of ' abab,' the New Hebridean helps to explain the Tasmanian and Australian demonstratives and pronouns above particularised. It suggests the derivation of Australian ' kodja,' ' kootya,' *tomahawk*, from ' kut ia,' *to cut*. It shows that the Australian forms 'murri,' 'main,' 'maar,' 'mail,' 'myal,' &c., equivalents of *man*, are variants of the same etymon ; and, what is most important, that the radical meaning is *male*. This last explanation probably disposes of the supposition that the word for *man* in Australian dialects generally is a racial appellative, and favours the presumption that it should be taken as equivalent to *male*. We are also enabled both to analyse and derive the anomalous form ' namail' (*man*) used on the Dumaresque River, New South Wales, ' na-' being the definite article unconsciously retained.

The New Hebridean further serves to analyse the Tasmanian

and Australian 'mookaria' and variants. It also suggests the true meaning of the term for the youth after initiation to the status of manhood. Especially valuable is the analysis which it enables us to make of so common an Australian word as 'wenyo' or 'wendyo,' *where*, into the interrogative particle 'wa' or 'we' and 'to' v. *to stand*.

The above comparison, so far as it goes, favours the conclusion that the primæval occupants of Tasmania, Australia, and the New Hebrides were of the one race.

ETYMOLOGY.

The Australian languages having no literature, considerable difficulty is experienced in tracing the derivation of words which have undergone change. People who have known one dialect well have declared that Australian words could not be derived, and that to attempt derivation would be futile. Some writers have gone to the other extreme of connecting Australian languages with Aryan or Semitic by fixing upon mere casual coincidences.

Apart from such changes as may have been introduced unconsciously for euphony's sake, I am persuaded that, just as in European languages, so in Australian, every syllable has its value and every word its history. We are unable to trace dialects backward in time, but we can track the changes of words by finding the same vocable occurring in different dialects at different and graduated stages of its modification. Cognate dialects show well-marked differences in the regular omission, addition, or substitution of certain letters.

The following are among the more important letter changes. 'S' or 'z' in the islands of Torres Strait changes to palatal 'ch' on the mainland. An old Tasmanian 'l' has in many cases become 'y' on the continent. 'L,' 'r,' and 'n' are frequently interchanged. Initial 'k' interchanges with 'w,' and either is elided; 'k' occasionally becomes 't' and occasionally 'y.' 'K,' 'ng,' 'n,' or 'ñ' marks a series of changes; 'ng' or 'n' sometimes becomes 'm'; 'ch' changes to 'ty' and 't'; 'r' changes to 't' or 'd,' and sometimes to 'th'; 'p' is softened to 'm.' A diphthong indicates the omission of a consonant, mostly a liquid, between the vowels. A peculiar tendency is for vowels of the

' u ' class ('ū,' ' o,' ' ŭ ') to change gradually into those of the
' a ' class (' a,' ' e,' ' i ').

The examples which follow will illustrate the letter changes
instanced.   The section on numerals may also be consulted.

Verbs are usually formed, as in Tasmanian, by adding a
verbal termination to a root or stem.   The most prevalent
Tasmanian termination, ' gana,' is common in Australia, and is
especially noticeable in the Victorian region, where most verbs
end in ' kan,' ' ka,' or ' ki,' sometimes contracted to ' k.'

Spelling according to English values of letters often dis-
guises the relationship of words.   Monosyllables, unless in
pronouns or particles, are usually to be suspected as corrupt.
A remarkable example is the West Australian ' waitch,' already
explained as the contracted form of ' uroi-kaiza.'   The West
Australian ' web' would hardly, at first sight, be taken for a
variant of ' kooia,' the more general word for fish, but re-
membering how ' k' is softened to ' w,' and that ' kooia' is
sometimes spelt ' queea,' and that the corresponding word in the
north-east is ' weenburra,' the origin of ' web' in some such term as
' wianbu' is highly probable.   Certain groups of objects are
designated by the same term.   Thus *wood*, *fire*, and some weapon,
usually the *spear*, are thus designated.   The same remark applies
to *sun*, *moon*, and *fire*, *sun*, *moon*, and *eye*, *stone* and *mountain*,
bark, *skin*, and *canoe*, *stone* and *egg*, *stone* and *tooth*, *egg* and *head*
(or skull), and probably *man* and *kangaroo*.   The name for
kangaroo in one dialect occasionally corresponds to the name
for man in a dialect adjoining, the reason being that both were
called *male*.

In the formation of compound words the Tasmanian practice
obtains on the continent.   A common name stands first and is
joined to a qualifying adjective.   I believe this principle of
nomenclature will explain nearly all the words of more than two
syllables, of course leaving onomatopœic words out of account.
The ideas which the names suggest were called up by features
in animals or things which would most strike the imagination.   And these ideas are singularly rudimentary in the
names of animals, being associated with the size, colour, or
shape of some member of the body.   In words, the significance
of which is now unknown to the aborigines themselves, we
perceive strange gleams of thought which have struggled down

the ages like light from faint stars in the unfathomable deep of heaven.

The derivation of Victorian 'kaiap,' *one*, has already been traced to a form like 'kurupana,' this being divisible into two equal parts. The first part, 'kuru,' or 'kura,' is the common type for *one*. It appears in 'kain' of West Australia, the 'n' corresponding to a final 'na' or 'nu' to give substantive value. In connection with this root I would direct attention to West Australian 'mau,' *three*, which is a contraction for 'mankura,' *three*, signifying literally *by-one* or *and-one*, being the latter part of the series 2 + 1.

'Miowera' is one of those soft, musical words which the colonists select as names for houses or ships. It is the Gippsland word for *emu*. 'Mio-' is contracted from 'murrio-' (as in Sydney 'marayong'), and this again is a corruption of 'nguruin,' which is derived from 'kuroi,' already explained as meaning *bird*. 'Wera' is an adjective the meaning of which I do not know; by analogy it probably means *large*. 'Koonawara,' a euphonious Victorian name for *black swan*, is formed of 'koonna,' *neck*, and 'wara,' *crooked*.

'Koondooloo,' a common name for *emu* in the north-east of the continent, is probably derived from 'koonna,' *neck*, and 'dooloo,' *wood* or *tree*, the idea suggested being that the bird was tall like a sapling. 'Kangaroo,' or, as originally spelt, 'kanguru,' now apparently obsolete at Endeavour River, is probably from 'ka' or 'kuggan,' *nose* or *head*, and 'guru,' *long*. The same name is applied in the east to the laughing-jackass, in such a form as 'kagurrin,' because of its big head or long beak. 'Wagan,' 'wakala,' 'workulla,' &c., variants of the commonest term for *crow*, are probably from 'wuro,' 'kala,' or 'kurla,' meaning *bird, wooden*, or *firestick*. This connection of the colour of the bird with fire probably accounts for the myths about the crows having the secret of making fire, and their stealing the fire, and so on. Another common name for crow, exemplified in 'wardung,' West Australia, 'woterkan' in Queensland, is probably from 'wuro,' *bird*, and 'tarkoo,' *black*. Strangely enough, the most prevalent name for the boomerang, extending from Cape York Peninsula to Melbourne in the east, and, as I believe, to the extreme south-west in the contracted form 'kaili,' appears to be identical with the name for *crow*;

whether the boomerang was named after the crow, or just named *wooden bird*, is not clear. The word 'bomrang,' in south of Queensland 'bobran,' is probably connected with 'boran,' *wind*, and has been named from the rushing noise it makes. Another derivation is from a New South Wales word, 'bargan,' meaning *crooked*. This is certainly the name in some places, but it does not correspond to 'bomrang.'

These observations will show that a great deal may be done in the way of tracing Australian words to their source, and I hope the principles here enunciated will prove serviceable to other investigators.

WORDS.

It is a common assertion that in Australian languages single words carry nothing in themselves whereby they may be distinguished as parts of speech. This rough generalisation is not absolutely correct. In Kabi, for instance, there are several special adjectival terminations which mark the adjective, and which in some cases have become inseparably affixed. And there is in most languages a distinct verbal sign, which in some of them is as much incorporated in the word as '-are' in Latin 'amare.' As a general rule, likewise, interrogative pronouns and interrogative adverbs are marked, if not by a peculiar termination, at any rate by a distinctive initial syllable.

There are numerous particles employed in various ways and positions with modifying force, *e.g.*, the word 'kna,' used in West Australia as the last word in a sentence to ask a question, with a value something like the English *eh?* The vocable 'inga' serves the same purpose in the south of Queensland. Compare also the particle 'ya' used in the Kabi dialect of the extreme east with a meaning like *come now!* Besides floating independent particles of this class there are those attached sometimes to a phrase to give it a substantive value, like the word 'midde' in Western Australia, meaning the agent. It is said that all verbs may be rendered nouns by the addition of this particle. Of this more fixed and dependent class the modifying syllables attached to verbs may be cited as examples. These are probably words like 'midde,' but of which the identity and original sense are in all stages of evanescence.

General consent denies an article, properly so called, to

Australian languages. Doubtless the numeral *one* and the demonstratives *this* or *that* can be filled in to satisfy a demand for an equivalent to *a* or *the*. And Mr. Thomas [*] gives an enclitic '-arter' and '-o' as elegant definitives for *the*. But these probably mean more. There is one feature of certain dialects, however, to which I would call attention. It is the disposition to elide an initial consonant which may possibly be explained as indicating the existence in Australia of what has been called in other languages, such as the Malay, *the unconscious article*. The best examples of this are to be found in the most central dialects which we have any knowledge of, vocabularies of which are given by Mr. E. M. Curr,[†] received from Alice Springs and Charlotte Waters Telegraph Stations, and from the Macumba River. There is a similar peculiarity in a dialect of the Palmer River,[‡] but there it is excessive. At Alice Springs such words occur as 'arkoppita,' *head*, 'ulgana,' *eye*, 'iniga,' *foot*.

THE NOUN.

Number is rarely marked save by distinct words. There are, however, exceptions. In the speech of the Narrinyeri (South Australia) the plural is indicated by a special terminal inflection, *e.g.*, 'korni,' *a man*, dual 'kornengk,' *two men*, plural 'kornar,' *men*. In the verb, number receives no sound-mark. A fallacious notion which has been widely circulated may here be referred to with the view of exposing it. The Australians, it is said, have no general names, but only special terms. There are scores of words in every dialect testifying to the contrary. Take the Kabi dialect as an example: it has a general name for animal, man, tree, stone, creek, mountain, and so forth. The only grounds for the delusion referred to are the facts that some classes of objects have not been generalised and that there is a preference for the special distinctive name, even where a general one exists. Thus, instead of speaking of a tree, the native prefers to specialise the particular kind of tree.

Gender is commonly distinguished by the addition of a word signifying male, female, man, mother, or the like, but, in special

---

[*] Mr. R. Brough Smyth's "The Aborigines of Victoria," vol. ii. p. 118.
[†] Curr's "The Australian Race," vol. i. pp. 412, 425.
[‡] *Ibid.* vol. ii. p. 398.

classes of words such as the phratric names, there are occasionally terminations distinctive of sex, as, for instance, about
Brisbane, Queensland, 'barang,' a male of the class 'barang,'
'baranggan,' a female of the same class.   But the most striking
case of phonic indication of gender comes from the Daly River.
I am sorry to be unable to give my informant's name, as my
information came indirectly, but I believe he is a member of
the Roman Catholic Mission at that place, and I hope he will
publish a memoir upon the very interesting dialect of which I
have received a sketch and a vocabulary.   In the dialect referred
to, which is known as the Daktyerat and is spoken on the left
bank of the Daly River, Northern Territory, four genders are
distinguished in nouns, adjectives, and verbs, viz., masculine,
feminine, neuter, and common, or organic and inorganic, the
general distinctive marks being 'y,' 'n,' 'w,' and 'm,' respectively, with sometimes a following vowel, and these inflexions
are initial in adjectives, *e.g.*, 'yidello,' *a big (man)*, 'nudello,' *a
big (woman)*, 'wudello,' *a big (thing)*, sex not distinguished,
'mudello,' *a big (object of any gender)*.   These marks are probably the consonantal radical of the third personal pronouns.
In all the languages of more elaborate structure the noun is
exceedingly rich in cases, and as a rule where these are said not
to exist the fact is that they have not been recognised.   The
cases comprise such as nominative, genitive, dative, accusative,
ablative (instrumental), abessive, adessive, commorative, locative
(with distinctions of *in, towards, from*).

THE ADJECTIVE.

The adjective is usually compared by supplying an adverbial
word with the sense of *very*; frequently comparison is effected
by reduplication, complete or partial, the superlative being sometimes marked by a reiteration of the duplicated syllable; cf.
'worbrinun,' *tired*; 'worbrinunun,' *very tired*; 'worbrinununun,'
excessively tired—regularly done.*   This hanging on to a letter or
syllable also implies continuity or intensity in the meaning of
the verb in some dialects.   Another mode of comparison in
adjectives is by singling out that object which surpasses the

* Mode used by the Melbourne blacks, *vide* R. Brough Smyth's " The
Aborigines of Victoria," vol. ii. p. 118.

other or others, and saying 'this big,' 'this good,' and so on. In opposition to the view that a word may be a noun or adjective indifferently by tacking on or omitting the case endings, and that there is no difference in form, I repeat what I have already remarked, that this is not invariably true, that, for instance, there are certain recognisable adjectival terminations, such as '-ngur' in the Kabi dialect, although they are affixed to only a limited number of words. This, however, is to be observed, that in Kabi, nouns may become adjectives by the addition of '-ngur' just as in English by affixing -like, in German -ig or -lich.

NUMERALS.

The Australian aborigines are singularly behind most other races in their numeral system and the general practice of arithmetic. The binary system of numbering is almost universal among the tribes. This is their basis of enumeration. The simple method of reckoning by the use of only two distinct terms is varied in some dialects by the possession of a distinct word for *three*, in which instances the method is not regularly trinary, but an irregular combination of the three basal forms to represent higher numbers. It should be noted that the seemingly distinct terms for *three* are analysable into the terms for $2 + 1$ occurring in some locality or another, or are the scarcely recognisable remains of such a combination. For instance, at Lake Amadeus the numerals run 'goochagoora,' *one*, 'godarra,' *two*, 'munkuripa,' *three*. 'Munkuripa' is compounded of 'mun' = *and* or *by*, and 'kuripa,' a variant of the commonest form for *one* in the east and south of Australia, appearing in such forms as 'koorbno,' 'kutupon,' 'kaiap.' So that the full form for *three* at Lake Amadeus would have been 'godarra mun kuripa,' unless we suppose that 'munkuripa' was adopted from another dialect in which 'kuripa' meant *one*, which is quite possible.

We discover by analysis that prior to the binary system there was a mode of reckoning by simply repeating the unit as often as required and coupling the terms by a conjunction thus, $1, 1 + 1, 1 + 1 + 1$, and so on. This accounts for the same expression being used for *two* in one place and for *three* in another. In process of time one of the terms dropped out of the combination for *two*, the conjunction and the remaining term became fused and sometimes worn down. The unit of one locality being employed to express a higher number elsewhere is occasionally due, simply to numerical indefiniteness. Compare 'kourapong'=4, used at Mount Rouse, Victoria; primarily it meant *one*. Compare also 'warpool' at Bloomfield Valley, Queensland, which, though obviously originally meaning *one*, and still used for *one* in other dialects, means there *five, ten, many*.

In a simple binary system, such as the Kabi of Queensland, the reckoning proceeds thus: 'Kalim,' *one*, 'boolla,' *two*, 'boolla-kalim,' *three*, 'boolla-boolla' or 'boolla-kira-boolla,' *four*, 'boolla-boolla-kalim,' *five*. On reaching this height of reckoning the aboriginal brain usually gets puzzled, and all higher numbers are named by such a term as 'kurwunda' or 'bonggan,' which, in that particular dialect, means *many*.

In various tribes the terms for hand and foot are introduced for purposes of enumeration. Thus in the Watty tribe, on the Murray River, 'kyup' (*kaiup*) 'murnangin,' *i.e., one hand*, means *five*, and 'polite' (*polait*) 'murnangin,' *i.e., two hands*, means *ten*.* Some tribes have reached a stage of decimal enumeration built upon the binary, and as a further development vigesimal; in both cases the terms are those for hands and feet.

So far as the evidence available goes, with the Tasmanians the lower numerals were not combined to form the higher, each number being represented by a distinct term. The following conclusions are deducible from the examples preserved, which vary considerably. The syllable 'wa' is a common terminal mark. The stem of the term for *one* is 'mara'; that of the term for *two* 'poua' or 'pia.' The termination '-la' or '-lia' seems to have been sometimes substituted for '-wa,' giving forms like 'boula,' or possibly '-la' or '-lia' may have been added to such a form as 'poua,' giving 'pouala' or 'poualia.'

The Tasmanian stem 'mara' has survived in various places on the continent as the equivalent of *one*. It assumes the form 'mal' or 'marl' in the Kamilroi dialect of New South Wales, 'moar' between the Leichhardt and Gregory Rivers, near the Gulf of Carpentaria; 'mirina' or 'murina' on the Lower Barcoo, a form which interchanges with 'matina.' The

* Beveridge's "Aborigines of Victoria and Riverina," p. 175.

same Tasmanian stem appears in 'barkoola-marna' (2 + 1), *three*, Lower Diamantina River, 'barkoola-matina' (2 + 1), Lower Barcoo River, 'boolar-martung,' (2 + 1), Moneroo, New South Wales; also in 'polimea' (2 + mea) *three*, Lake Condah, Victoria.

The other Australian numeral forms not derivable from the old Tasmanian are reducible to a very small number of bases. In addition to 'mara,' already dealt with, the distinct terms for *one* are chiefly:—

'Kuru-po-na,' the most prevalent. It is not found exactly in this form, but the immense diversity of variants points to such an original. One of the nearest approaches to the typos is 'koorbno,' occurring on the Diamantina River, Queensland. Thence north-eastward to the coast it is worn down to 'nupoon,' 'nobin,' and similar forms. Apparently from a focus at the head of the said river this numeral has been carried south-westward to the most south-westerly point of Australia, southward to the most southerly point in Victoria, and south-eastward to the extreme east of the continent. The changes which it has undergone in transit are still traceable, and will be sufficiently exhibited on the subjoined table. Some of the extreme variations, like 'kaiap' of Victoria, are scarcely recognisable. A very important variant has made its way along the east of Cape York Peninsula, and has reached as far south as the tropic of Capricorn. It assumes such forms as 'warpa' (Port Denison), 'wirburra,' Belyando River, &c.; we pick it up again in Torres Strait as 'warapune' (Prince of Wales Island), 'woorapoo' (Warrior Island), 'nrapon' (Saibai Island). It is thus traced to within two miles of the New Guinea coast.

In New Guinea again there are other variants, such as 'koapuna' (Bula'a), 'abuna' (Aroma), &c. (see table). This, the most common term for *one*, and one of the most ancient, is not of Tasmanian origin. It has been introduced from New Guinea by the second tide of human life which overspread Australia, and yet it is distinct from both the Melanesian and Polynesian numeral types.

As regards derivation, it seems composed of two radicals, represented by 'kuru' or 'kura,' and 'pa' or 'po' respectively. The former appears in West Australian 'kain,' the '-n' being for '-na' or '-nu,' probably to give substantival power, just as the whole word 'kurupo' takes on '-n' or '-na' for the same purpose.   What the primitive meaning was would be difficult, if not impossible, to determine.   There are hints that the roots had demonstrative force ; thus, in the Kabi dialect of south-east Queensland 'karva' means *other* (Lat. *alius*), it is modified 'karvana' or 'karvano,' and 'v' in Kabi stands for 'b' in other dialects.   The local equivalent at the Macdonnell Ranges in the heart of Australia is 'arbuna,' *another*.   'Karba' is also used there.   The syllable '-pa' I regard as a demonstrative corresponding to the third personal pronoun singular (with 'parna' plural), Adelaide.   'Panna' is third singular in Parnkalla dialect.

'Kula,' 'kuala,' 'kualim,' &c., another typical term for one, is, I feel sure, simply a variant of 'kura.'   In the north-west of New South Wales it occurs simply. · In the same neighbourhood and generally in the east central belt it occurs in 'barkoola,' *two*, where 'bar' has some such meaning as *and* or *another ;* compare Victorian numerals in which 'ba' or 'pa' joins *one* to a number preceding.   'Barkolo' or 'parkula' also means *three* in South Australia, according to a principle already laid down, viz., that in the higher numbers the word for *one* came last, and when a new term for *one* was introduced the older term got the higher value.   At Port Darwin 'kula' occurs as part of 'kulagook,' *one*. At Hawkesbury River, New South Wales, it may form part of 'workul,' *one*, and at Omeo, Victoria, part of 'warkolala,' *two*. The identity of 'kula' with 'kura' is borne out by its occurring in such words as 'koolbarro,' *three*, which is the local form at the Upper Burdekin corresponding to 'koorborra,' *three*, at places not far distant, the latter being a variant of 'kurupana.'

'Wongara,' 'ungar,' 'onegan,' 'ungal,' 'yongole,' &c.   This seems connected with preceding, but not distinctly.   Its range is comparatively limited.   It is doubtless a later term.   Its New Guinea equivalent is 'aungao.'   It appears to have been brought to Australian shores about the neighbourhood of Hinchinbrook Island in the form of 'yongool.'   At Broadsound, some 250 miles to the south-east, it occurs as 'onegan' (*i.e.,* 'wongan '). Inland it has passed in a southerly direction to the Paroo and Maranoa Rivers, a distance of some 600 miles.   Westward it has reached the mouth of the Leichhardt River.   It may be the same vocable which prevailed at Sydney and neighbourhood in forms like 'workul,' 'wogul,' &c. ; at Mount Gambier, South

Australia, in the form 'wondo,' and at Woodford, Victoria, as 'waando.' This view is supported by the terms for *three* at the last two places, viz., 'waawong' and 'wrow-wong' respectively. There is a presumption in favour of its original application being demonstrative. In the Woolna dialect 'wongalyer' (stem 'wonga' or 'wongal') means *that*, and on the Cloncurry River 'wallegul' means *this side*.

Two other terms for *one*, 'kutia' and 'whaityin,' I regard as variants of one radical form. I shall indicate their distribution separately.

'Kutia' is scattered widely, being most persistent in the extreme west. 'Guddee,' on the Upper Murray, is probably the local homologue. 'Kudjua' occurs at the mouth of the Burdekin, Queensland, for *three*, 'kutchoo' is *three* further north at the Thornborough diggings, 'marukutye' is *three* at Adelaide. The two former are remnants of a combination, 2 + 1, and '-kutye' in the Adelaide word doubtless stands for *one*, being represented in the terminal syllables of 'yammalaitye' (1) and 'parlaitye' (2). It is therefore also represented in Victorian 'polaitch' (2) which in structure corresponds to 'barkula' used elsewhere. Strange to say, 'goochagoora,' *one*, at Lake Amadeus combines this term with the more primitive 'kura.'

'Whychen' or 'whaityin,' with variants 'wigin,' 'wogin,' 'watchin,' &c., is more restricted. This word reached Australia about the mouth of the Burdekin. It is not found north of Townsville. Its homologue occurs in the form 'koitan' on Woodlark Island, to the south-east of New Guinea. It has penetrated inland in the forms 'wogin,' Belyando River; 'itcha' on the Warrego and Paroo. On the Darling it has displaced the more ancient 'koola,' and appears there as 'neecha,' 'ngitya.' Its furthest south is Piangil, on the Murray, where the form is 'yaitna,' its occurrence here, however, compels us to identify it with the 'aitch' in Victorian 'polaitch' and consequently with 'kutia.' 'Waityu' is the synonym at Cooper's Creek, 'waityuali,' near there, on the Wilson River. 'Ninta,' in the very heart of Australia, is probably the local corresponding form.

The above three types (regarding 'kura' and 'kula' as identical, as also 'kutia' and 'whaityin') leave very few terms for *one* unrelated.

The etymons for *two* are fewer than those for *one*. Four types especially prevail—' boolla' along the eastern watershed and headwaters of the Barcoo and Darling Rivers, extending southward from the Burdekin; 'barkoola' from watershed of Gulf rivers over the east-central district; 'polaitye' in Victoria generally and south of South Australia; and 'kootara' in the west. But these are reducible to three inasmuch as 'boolla' is certainly a contraction of 'parakula.' 'Kula' occurs for *one* only in the north-west corner of New South Wales along with ' boolla' *two*, and for many miles round in all directions the terms for *two* (occasionally *three*) are such as 'barkula,' 'parakula,' 'piakula,' &c.

' Barkoola,' 'rankool' (Murray River), 'parakula.' 'Bar-' is most probably a conjunction signifying *and*. The full original form of this numeral would be 'koola-para-koola,' 1 + 1. This supposition accounts also for its occurrence as *three*. At Boolcoomata, where ' barkoola' is *two*, 'koola' is *one*.

' Polaitye,' 'polaitch,' 'poolet,' &c., Victoria; ' parlaitye,' (Adelaide, S.A.), I once thought to be variants of ' boolla,' but am now convinced that they are analysable into 'para' 'aitye' (= *and one*) after the manner of 'parakula.'

' Kootara,' 'koodthera,' 'koochal,' 'kujara,' are the prevailing western terms. In the centre this type occurs with 'koo,' the numeral index (as we may call it), omitted, as 'tera' at Macdonnell Ranges, and 'trumma,' &c., in neighbourhood.

Such forms as 'ooroopoochama,' *three*, in centre of Australia, are easily explicable as '(k)ooroopoo-trama,' 1 + 2. The homologues of ' kootara' appear in Torres Strait as 'quassur' (Prince of Wales Island), and ' ukasara,' Saibai Island.

Having already enunciated the principles upon which the terms for *three* and higher numbers have been formed, further analysis is rendered unnecessary.

The subjoined table will show the changes which several types have undergone, how they may be traced along lines converging in the north-east of Australia, and that homologous terms are to be found in New Guinea or adjacent islands.

AUSTRALIAN AND NEW GUINEA NUMERALS COMPARED.

New Guinea Dialects	Torres St., Saibai	Bula'a 300m. N.E. of Saibai	400m. E. by S. from Saibai Kerepunu and Aroma	600m. E. of Saibai Woodlark I.	Inland, nr. Mt. Yule Upper St. Joseph
Terms for ONE	Urapon	ka or koapuna	obuna abuna	koitan	aungao
Australian Equivalents / Australian Dialect	Warpur MACKAY, Q.	koorbno DIAMANTINA, Q.	kutupona GIPPSLAND	nupoon nobin nuboon E. COAST C. YORK PEN. 16° S. to 18° S.	whychen CAPE R., Q.	wonga BURDEKIN

TABLE SHOWING AUSTRALIAN VARIANTS OF NEW GUINEA FORMS.

SAIBAI, N. G.   URAPON.

Warrior I., Torres St.	woorapoo
P. of Wales I., Torres St.	warapune
Port Denison, Q.	warpa
Belyando R., Q.	wirburra
Port Mackay, Q.	warpur
Rockhampton, Q.	werpa
Peak Downs, Q. (about 23½° S.)	woorba

SAIBAI, N. G.   UKASARA (two)

Prince of Wales I.	quassur
Port Denison	kotoo
Flinders and Cloncurry	kurto
Alice Springs, C. A.	trumma
Macdonnell Ranges, C. A.	tera
Nickol Bay, W. A.	kootara
Murchison R., W. A.	kootera
St. George's Sound, W.A.	koochal

NEW GUINEA.   KOAPUNA.

Diamantina, Q.	koorborra koorbno
Cooper's Creek, S. A.	koornoo
E. of Lake Eyre	koornoo
N. E. of Lake Eyre	koono
N. Coast of Great Australian Bight, W. A.	kyunoo
W. Coast of ditto	kean

S. W. Corner of Australia	kain
E. of Lake Torrens	koobmana
E. of Great Australian Bight	kooma
Castlereagh R., N.S.W.	ngunbeer
Wellington, "	oonboyie
Waljeers, "	koolnebine

Bumbang, V.	geyabi
Victoria (general)	kaiap
Gladstone, Q.	karoon
Burnett R., Q.	karboon
Stradbroke I., Q.	kurraboo
Gippsland, V.	kutupona
Melbourne, V.	karnboo
Mount Rouse, V.	konrapong = 4
Streaky Bay, S. A.	karboo = 3

WOODLARK I., E. OF N. G.   KOITAN.

Cape R., Q.	wigin, whychen, witchen
Belyando, Q.	wogin
Paroo and Warrego, Q.	itcha

Darling, N.S.W.	neecha, ngitya
Piangil, V.	yaitna
Cooper's Creek, S. A.	waityu
Central Australia	ninta
West Australia	kooten

NEW GUINEA.   AUNGAO.

Upper Flinders, Q.	ungar
Mount Black, Q.	anga
Paroo, Q.	wongara
Maranoa, Q.	wongara
Hawkesbury, N. S. W.	workul
Sydney, N. S. W.	wogul
Mount Gambia, S. A.	wondo
Woodford, V.	waando

THE PRONOUN.

The pronouns are specially remarkable for the almost universal currency. of certain forms, both of stem and (less uniformly) of case-ending, notably those of the first and second persons singular. The first and second persons singular are generally of the central Indian 'nan-nin' type ('ngan-ngin' rather in Australia); in some cases the plural has the same base as the singular, with generally a syllable marked by the letter 'l' to indicate plurality, this also being an Indian feature. In the first and second persons there is usually a dual, the first dual being, sometimes at least, such a compound as 'we-thou'; the second sometimes is, sometimes contains, the numeral *two*; occasionally a trial number for the first person is met with and a dual for the third. In the West Australian speech different pairs are indicated by different details in the three persons, significant of such relations as (1) husband and wife or people greatly attached; (2) parent and child, uncle and nephew, and the like; (3) brother and sister, or a pair of friends. There is usually no phonic connection between the third persons singular and plural, a common form of the third plural has the etymon 'than' or 'tin'; the distinction of sex is not usually marked in the pronoun, though there are exceptions. Decayed pronouns are frequently incorporated with nouns to indicate possession, in such forms as *father-my, father-your*, and also with verbs as the personal index not yet assimilated so as to obliterate the origin, and in such cases the position is usually terminal, though here again the case of Daktyerat dialect is a clear exception, where the pronominal element may be either initial or medial. The pronoun is also well supplied with cases, and possessive forms are in some dialects capable of declension like nouns. Demonstrative pronouns are also declinable like the personal in certain dialects, as, for example, that of Lake Macquarie in New South Wales. The interrogative pronouns and interrogative adverbs may be mentioned together as having much in common. The radical elements are usually 'ngan' in *who* or *what*, 'wendyo' in *where, when,* &c., 'min' in *how, why, how many, what*, &c. These are declined to correspond to a variety of shades of meaning, and they are among the most uniform and widespread words.

TABLE OF PRONOUNS.

| ENGLISH | TORRES STRAIT, NEAR NEW GUINEA | | QUEENSLAND | | | N. S. WALES | S. AUSTRALIA | W. AUSTRALIA |
	SAIBAI	KOWRAREGA	BLOOMFIELD VALLEY	KABI	TURRUBUL	KAMILROI	DIYERI	
I	gnai	ngai, ngatu	aio	ngai, ngadhu	ngai	ngaia	athu	nganya
my . . .	gnau	—	aiko	nganyunggai	—	ngai	ani, ni	nganaluk
thou . .	gni or gnido	ngi, ngidu	youndo	ngin, ngindu	nginta	nginda	yondru	nginni, nundu
thy . . .	gninu	—	youno	nginyonggai	—	nginnu	—	nunoluk
he or she .	gnoi	nuo nudu	nulu	ngunda	wunnal	ngerma	noalia	bal
his . . .	gnugnu	—	ngongo	ngundano	—	ngundi and ngerngu	noonkani	baluk
DUAL								
we two .	gnabugnaba	albei	ali	ngalin ngin	ngullin	ngulle	uldra & yanna	ngalli, &c.
you two .	gnipel	ngipel	youbal	bula	ngilpung	ngindale	yula	—
TRIAL								
we three .	gnalpa gnalpa	—	ana	—	—	—	—	ngalata
PLURAL								
we . . .	gnapamura	arri	angin	ngali or ngalin	ngulle	ngeane	ali, yana	ngannil
you . .	gnitamura	ngitna	yourer	ngupu	ngilpulla	—	—	nurang
they . .	tana	tana	tanner	dhinabu	ngarma	—	thana	balgun

Besides the paradigms which are given in the treatment of separate dialects I have prepared a table showing the substantial unity of the Australian pronoun. The essential features are observable in the dialect of Saibai, two miles from the New Guinea coast, and understood on the coast. This, to me, is suggestive of the route along which the pronoun was conveyed. The double subject is noticeable in the Torres Strait singular forms, the one in 'du' or 'tu' indicating action. There is a considerable amount of agreement in the characteristic vowels of the singular, *a* for the first person, *i* for the second, *u* for the third.

Regarding the radical import of the pronouns a little may be said. Dr. Latham in a note at the close of McGillivray's "Voyage of the *Rattlesnake*" connects them with the demonstratives. Whether the first and second persons may be thus connected I shall not venture to conjecture. But it is interesting to observe that the third person singular of Kamilroi 'ngerma' (stem 'nger') and the third plural of Turrubul 'ngarma' are practically identical with the old Tasmanian word 'nara,' *he, they,* and *that.* A reference to the Comparative Table will show that 'nara' is also the equivalent for *they* at Nguna, New Hebrides. The very widely prevailing radical for *they,* viz., 'tana,' was also, there is good reason to believe, a demonstrative. In the sketch of the dialect of the Macdonnell Ranges which follows, 'tana' is the stem of the demonstrative *that,* as 'nana' is the corresponding stem for *this.* 'Thana' occurs as the distinctive mark of the third plural in the Yarra River (Melbourne) dialect, its identity being obscured by its being attached to the peculiar pronominal sign 'moromba-.'

The common second person singular 'ngin,' 'ngindu,' may be a survival of the Tasmanian 'nina,' 'neeto,' *thou.*

PREPOSITIONS AND CONJUNCTIONS.

Prepositions can scarcely be said to exist except in the sense of preformatives, and where they are represented as being in use, as, for instance, in the contribution to Mr. Brough Smyth's work of the language spoken at Lake Tyers, Gippsland, the statement is liable to arouse the suspicion that the idea of separate prepositions may be due to a straining to conform the native speech to European types.

The conjunctions are few and connectives rarely employed, but occasionally adverbs are very numerous, and they appear in some instances, as in Kabi, to be formed from nouns after the manner of the formation of adjectives already cited, but with a peculiar adverbial ending. In the dialects of central Australia, and to a large extent elsewhere, the use of adverbs is superseded by the abundant modifications of the verb.

THE VERB.

The verb is the most complex and elaborate of all the parts of speech. So numerous and diverse are its modifications that they astonish and perplex the student and puzzle him in his endeavour to designate and classify. In fact, European grammatical terms will not embrace them in their rich abundance, or, to vary the figure, are insufficient to clothe them. We generally find English writers conjugating Australian verbs after the approved methods of English grammar, which is like an attempt to spell all the words in the English language by the use of only half a dozen letters. The verb has what may be called *forms*, such as simple, reflexive, reciprocal, and there are several *moods*, as optative, inceptive, infinitive, imperative. Then there are *tenses* in considerable variety, as for instance in the Wiradhuri, indefinite present, definite present, aorist, indefinite perfect, definite perfect, to-day's perfect, yesterday's perfect, distant perfect, and so on with future, until, according to Dr. F. Müller, fourteen tenses are enumerated.*

This is, of course, a description of the most complex types, but the languages generally have such *forms, moods, tenses*, and *participles* besides, although in a great many dialects the number of changes is much more limited than in the Wiradhuri. And it is just possible that some writers have needlessly multiplied forms by incorporating with the verb adverbs which should have been regarded as separate words. A caution must be given against supposing that the verbs are generally regular, so far as my personal experience goes, which is confined to the

* Some dialects have an active and passive voice, as the Lake Macquarie, in which an incorporated pronominal element in the nominative indicates the active, a similar element in the accusative marks the passive.

knowledge of one dialect *acquired directly from the blacks*, there
is great irregularity and many verbs are defective. The position
of words in the sentence is subject to considerable variation
according to dialect. Commonly in categorical sentences the
nominative comes first and is immediately followed by the object ;
qualifying words if present succeeding their respective subjects ;
after the object comes the adverb and finally the verb.

CHAPTER XIII

OUTLINES OF GRAMMAR

Grammatical sketch of Tasmanian and of five Australian dialects representing the linguistic classes—Tasmanian—Wimmera, Victoria—Kabi, Queensland—Specimen in Kabi, with translation—West Australia—Diyeri, South Australia—Macdonnell Ranges, Central Australia.

I NOW submit a brief outline of the grammatical forms of six different languages, furnishing an example (not necessarily typical) for each of the six classes into which I have divided the languages of Australia.   Absolute consistency in spelling native names must not be expected.

THE LANGUAGE OF THE TASMANIAN ABORIGINES.

Authorities : The vocabularies and phrases in E. M. Curr's " The Australian Race," vol. iii. pp. 604-672.

The language of the Tasmanian aborigines became extinct with Truganini, said to have been the last of her race, who died in 1876.   Several vocabularies have been preserved, but most of them are very brief.   No attempt appears to have been made to master and place on record the grammatical structure of the language, and the dialogues and phrases that have been handed down afford but scanty material for the deduction of general principles.   So little has been done in this way and so little data are to hand for generalisations that it might be considered hardly worth troubling to attempt to arrive at any order.   But apart from the mere interest attaching to any vehicle of human thought there are certain special features about the Tasmanian language that might dispose to a close study of it.   First of all it is the language of the advance-guard of the human race in the Asiatic hemisphere, and has probably not been much affected by the introduction of foreign elements. And then further, it forms, according to the present writer's view, the substratum of the Australian languages generally, so that if they are to be studied with any degree of comprehensiveness the influence of Tasmanian speech upon them must never be left out of account. A difficulty has been experienced in finding any relationship between Tasmanian and Australian dialects; one reason for this difficulty has been the assumption that Tasmanian should form a kind of Australian dialect. It is not, strictly speaking, an Australian dialect at all, but a distinct language, the language of the real Australian aborigines, but in modern times not found in Australia except as a barely recognisable ground-colour of most Australian dialects and more decidedly of those of North Australia, West Australia, Riverina, and Victoria.

PHONIC ELEMENTS.

Vowels.

a			
e	ŭ	o	ō
i	ī		ñ

Diphthongs.

ai   au   ei   iu   oa   oi

Consonants.

k	g		ng			
t	d	th	y	l	r	n
z	sh	j palatal				
p	b		w	m		

In Peron's list 'h' and 's' also occur.

There is a very decided preference for initial consonants and terminal vowels, the prevailing terminal letter being 'a.' A few words terminate in 'n,' 'r,' 'l,' or 'k.' Final 'k' is specially a mark of the adjective in the southern dialect. 'L' and 'r' occur frequently as initials, a mark also of the Victorian and Northern Territory dialects on the mainland. 'Ng' occurs both initially and in other positions. Such combinations as 'kr,' 'gr,' 'tr,' 'dr,' 'pr,' 'br' are common, but there are clear indications that originally a vowel intervened—*e.g.*, 'prugga' and 'parugga,' *breast*.

A comparison of the local variations of words leads to the

conclusion that in the west of the island the language was more decayed phonetically than in the other parts.

THE NOUN.

The Tasmanian agrees with the majority of Australian dialects in being without inflectional signs to mark number and gender in the common noun. The noun, as well as other parts of speech, is modified chiefly if not exclusively by post-positions. The terminations '-na' in the east and a corresponding '-lia' in the west are common nominal or definitive signs. From their being almost invariably affixed to names of organs of the body, it has been suggested that they may be pronominal affixes, but the fact that they frequently terminate other kinds of substantives is fatal to such a supposition.

As exemplifying the above remarks I may cite the terms for *man* and *woman*. In the east the word for *man* is 'pugga-na,' and in the west 'pa-lia.' 'Lowan-na' is eastern for *woman*, while the western is 'noa-lia.' The terminations are separable. Compare also 'pugga-luggan-na' (lit., *man-foot*), *footmark*, of the east with 'pa-lug' of the west.

It would be rash to attempt the formation of a paradigm of declension from the scanty material preserved in Dr. Milligan's dialogues. The examples are, however, sufficiently numerous to indicate the principles of construction. Nouns are modified by affixes as generally in Australian dialects. The sign of the dative is '-to,' '-ta' or 'tu'; thus, 'luna' is *house*, 'luna-tu,' *to the house*, 'nanga,' *father*, 'nanga-to,' *to (your) father*. A sign for the locative is '-reta,' *e.g.*, 'luna-reta,' *in the house*. 'Lia' is *water*, 'lia-titta,' *in the water*.

Pronominal suffixes are also employed. Thus, 'nanga-mea,' *my father*, 'nang-eena,' *thy father*, 'pugherá-nymee,' *his hair*. Where a pronominal and a case-modification are both present the former comes first, *e.g.*, 'luna-mea-ta,' *to my house*.

The language is partial to compound words, of which the constituent elements remain easily distinguishable, as for example :—

prugh-walla . . .	*breast-water*	*milk*
pugga-lee-na . .	*man-light (the)*	*the sun*
pugga-nubra . .	*man-eye*	*sun*
kul-lugga-na . .	*bird-foot (the)*	*talon*
mongta-lin-na . .	*eye-house (the)*	*eye-lash*
gooa-langta . . .	*bird-big*	*eagle*

The above method of word-formation is likewise characteristically Australian.

PRONOUN.

The information available regarding the pronouns is exceedingly meagre. The few authorities who have mentioned them agree pretty closely. The inflectional changes do not appear numerous. Exactly the same forms are given for subjects and objects. The possessive case follows the noun which it qualifies, or is postfixed to it in a contracted form. The pronoun is rarely expressed separately from the verb which by implication it governs. It may be expressed in the verbal form, but this is doubtful.

PRONOUNS.—*First Person.*

	Singular.	*Plural.*
Nom.	. . mana, meena	warrander
Gen.	. . mena, -mea	
Dat.	. . miape, mito	
Accus.	. . pawahi, meena	

Second Person.

	Singular.	*Plural.*
Nom.	. . nina, neeto	neena, nee, ninga
Gen.	. . -eena	
Accus.	. . neeto	

Third Person.

	Singular.	*Plural.*
Nom.	. . nara, narrar (*m.* and *f.*) ; niggur (*neut.*)	nara
		Accus. nara

Interrogatives.

wanarana, telingha, tebya, pallawaleh, tarraginna, *what.*

Demonstratives.

narrawa, *this* (*is*) ; avere, nara, *that.*

ADJECTIVE.

The adjective is generally indicated by the termination '-ak,' or '-iak,' especially in the east; '-é' or '-té' is a frequent adjectival termination in the south. Privative forms are distinguished by an affixed negative, as in the following words : 'lowa-timy,' *wifeless;* 'payea-timy,' *toothless;* 'pugga-timy,' *childless;* 'poruttye-mayek,' or 'paruye-noyemak,' or 'paroy-

time-na,' *leafless.* As on the Australian continent some words
have a wide range of application, thus, 'eleebana' is employed
in such senses as *good, beautiful, sweet, right, fragrant,* &c.

NUMERALS.

In the general introduction to the language the numerals
have already been noticed.

> marrawah; marai (P); marrarwan, borar, parmere (N); parmery (J);
>   *one.*
> piawah, pooalih, buwah; bura (P); boula, calabawa (J); pyaner-
>   barwar (N); kateboueve (L); *two.*
> lia winnawah, talleh; aliri (P); wyandirwar (N); *three.*
> pagunta, wullyawa; *four.*
> pugganna, marah, karde (G); *five.*

[The capitals stand for the authorities Peron, Norman,
Jorgenson, Lhotsky, Gaimard (in " Voyage de l'Astrolabe "); the
other terms are from Milligan.]

VERB.

The verbal termination is usually well marked. The following forms at least are determinable. On the east coast '-kuama,'
'-kena,' '-gena,' '-gŭna,' '-tone'; on the south, '-gana,'
'-gara,' '-bea,' '-tone'; on the north-west and west, '-bea.'
As illustrating the variety of termination the following
typical forms of one word will serve:—East 'ton-guama,' south
'ton-gane,' west and north-west 'tona-bea,' all signifying *to
gulp.*

ADVERB.

> namelah, nayeleh, wabbara, *when* and *where.*
> ungamlea   *where.*

THE LANGUAGE OF THE WIMMERA DISTRICT IN NORTH-WEST OF VICTORIA.

Authorities: Revs. F. A. Hagenauer, A. Hartmann, and F. W. Spieseke,
in contributions to Mr. R. Brough Smyth's "The Aborigines of Victoria,"
vol. ii. pp. 39, 50, 55, 76. The accounts of the two latter contributors fairly
agree. Mr. Hagenauer's shows considerable differences.

PHONIC ELEMENTS.

Vowels.

a ă ä
e ŭ ŏ o ŏ
i ī u û
diphthongs ai au oi

Consonants.
k g ch ng
t d ty or tch dy y r rr l n ñ
p b v w m

As in the N. S. Wales and S. Queensland division there is a marked preference for consonants at the beginning of words. Any consonant except ' r ' may be initial. There is no restriction as to terminal letters. We find here initial ' l ' and medial combinations as ' rt,' ' pkr,' ' rpk,' ' rmb,' which would not be tolerated in the dialects of Queensland and N. S. Wales.

THE NOUN.

Difference of number or gender is not marked by sound. For the plural, above a certain small number a term signifying *many* is added or the word is reduplicated. The noun is thus declined—

	Singular.	*Singular.*
Nom. . .	wūtye, *a man*	galk, *a stick*, wille, *opossum*
Gen. . .	wūtyūgitg	nom. agent willetch
Dat. . .	wūtyuk	
Acc. . .	wūtye	galka and galko
Abl. . .	wūtyūkal, *by a man*	
Exat. . .	wūtyenung, *from a man*	
Erg. . .	wūtyel, *with a man*	galko, willedyal, *in an opossum*

THE PRONOUN.

The pronoun shows considerable modifications. It is subject to be attracted to other parts of speech in abbreviated form, *e.g.*, the possessive pronoun is affixed to the noun ' mam,' *father*, thus—

mam-ek, *my father*	mam-endak, *our father*
mam-in, *thy father*	mam-angngodak, *your father*
mam-ūk, *his or her father*	mam-ennak, *their father*

Adjectives may become passive verbs by a similar process, thus ' katyelang,' *sick*, makes ' katyelang-au,' I *am sick*, ' katyelang-ar,' *thou art sick*, and so on. The importance of the pronominal element affixed to the verb will be observed further on. I shall show two tables of the pronouns, the first by Mr. Spieseke, the second by Mr. Hagenauer, as I think that both are required for an explanation of the verb and for a fuller view of the language.

Personal Pronouns—*First Person.*

	Singular.	*Plural.*
Nom.	ngan	ngo
Gen.	ngek	ngendak
Acc.	ngerrin	ngandank

Second Person.

	Singular.	*Plural.*
Nom.	ngar	ngat
Gen.	ngin	ngodak
Acc.	nganung	din

Third Person.

	Singular.	*Plural.*
Nom.	ngait	ngaty or ngatch
Gen.	nguk	ngeannak
Acc.	ngun	ngin

Second Table—*First Person.*

	Singular.	*Dual.*	*Plural.*
Nom.	walŭnek, ñanon	walŭnganŭk	walŭngingorak, ngarra
Acc.	walŭnŭngek	walŭngŭngnok	wallogingorak
Abl.	walŭgalik, *by me*	walŭngnŭngnalŭk	wallogaringorak

Mr. Hagenauer also gives a dative singular 'gangek,' *for me,* and a genitive plural 'gorak,' *ours.*

Second Person.

	Singular.	*Plural.*
Nom.	walŭngin	bŭlet wŭl
Acc.	walŭngin	bŭlet wŭl
Abl.	walŭgalet	bŭlet wŭlek
Voc.	walŭngin	bŭlet wŭlan

Third Person.

	Singular.	*Dual.*	*Plural.*
Nom.	gilla, ñogung, ño	bŭlang	giang

It will be observed that in the first and second persons in the second table there is an introductory particle, 'walŭ'; this probably is some word such as *self,* more exactly defining the pronoun; when this is decapitated the likeness of the two tables is rendered more close.

Interrogatives.

Nom.	wiñar wiñaru, *who*	wiña and gio, *where,* wiñang, *how far*
Gen.	wiñarait	wiñatuk, *which*
Dat.	wiñarangait	
Acc.	wiñer	ngan, *what*

ngak, *why,* ngango, *how*

ADJECTIVE.

The article is not present. The adjective does not seem to be declined. Comparison is denoted by reduplication. The adjective commonly precedes the noun, a somewhat rare position for it to occupy.

NUMERALS.

The numeral system is binary. The method of enumeration is 'kaiup,' *one*, 'būlet,' *two*, 'būlet kaiup' or 'būlet ba kaiup,' *three*, 'būlet būlet' or 'būlet ba būlet,' *four*, and so on. The natives of this tribe counted up to twenty, which is 'būletgedi mañya' (apparently *twice-two hands* or *both feet and hands*).

THE VERB.

The verb seems simpler than in most dialects, but the apparent simplicity may be due to want of full information. Conjugation is by post-positions. The pronoun abridged is attracted to form an affix, showing number and person, and a different fuller form (apparently an accusative case) of the pronoun distinguishes the passive from the active voice. The word 'mala' with the appearance of an auxiliary verb occurs along with the principal verb in perfect tenses and in the potential mood. Of the different writers one places it before the principal verb and joins on the pronominal affix to it, the others make it succeed the principal verb. Time seems hardly distinguished save by this word 'mala' with the force of *have* or *had*, and by a word such as 'maluk,' signifying by-and-by and denoting the future. Certainly in Mr. Spieseke's view of the verb the particle 'in' is introduced between the stem and the pronominal element to indicate past time, distinguishing the imperfect from the present; the same particle is affixed in Mr. Hartmann's view to mark the future. This double use raises distrust in its temporal power. There appear to be at least two participial forms, the imperfect ending in '-na,' the perfect in 'n' with a preceding vowel, as 'prinna' (*is*), *rising*, 'prinnon,' *risen.*

The following is a table of the pronominal elements used as post-positions to distinguish number and person in the verb, the first consonant of the affix in the active voice may be 'g,' 'y,' 'ng,' or may be elided as ease of utterance may require.

In the passive voice the pronominal element is free from accommodating phonic change, and by comparison with the declension of the personal pronouns it will be recognised as the accusative case. The third person singular of the verb in the present imperfect and future tenses is joined with the various accusatives to form the passive voice in these tenses, so that the present indicative passive would run *he sees me, he sees thee,* &c. This mode of forming the passive corresponds with that which prevailed at Lake Macquarie, New South Wales.

	ACTIVE VOICE.			PASSIVE VOICE.	
	Singular.	*Plural.*		*Singular.*	*Plural.*
1st Person .	-yan	-yango	(ñaing, *sees*)	-naingn	(ñaing, *sees*) -niyangoring
2nd ,, .	-yar	-yat	,,	-ñiurnung	,, -niyurding
3rd ,, .	-n, ng, or -kinya	-gitch or -gatch	,,	-ñitch	,, -nityaning

For imperfect tense of passive, 'nyain,' *he* or *they saw*, is used throughout; and for the future, 'ñakin,' *he will see*, followed in both cases by the pronominal affixes as above.

THE KABI LANGUAGE.

Authority, personal observation. A fuller but less systematic notice of this dialect was contributed by me to Mr. Curr's work, "The Australian Race,"* which would illustrate and support my remarks here. For two or three points the Rev. W. Ridley's account of Dippil is drawn upon.

Kabi is spoken chiefly in the basin of the Mary River. Queensland. The name is one of the negatives of the language. I have taken this dialect as a specimen of the elaborate dialects of the east, not because it is the most highly developed and richly modified, but because it belongs to that class, showing the various distinctive features of its near relatives the Kamilroi and Wiradhuri, and especially because rather than enter upon other men's labours I prefer, where possible, to tabulate a dialect which has not been systematically treated by any one else.

PHONIC ELEMENTS.

Vowels.

a ā ả

ę (as in *yet*, English) e ŭ ǫ (as in English *ton*) o ŏ
i ī u ū

Diphthongs.

au ai iu oi ou ua ui

Consonants.

k g   ng

t d th dh ty (almost like palatal ch) y r rr (muffled cerebral) l n ñ ndh
p b v   w m

Kabi has no words beginning with '1' or 'r,' and its terminal letters are '1,' 'm,' 'n,' 'r,' 'ng,' 'ndh,' and vowels. Initial vowels sometimes occur, but very rarely. There are occasionally as initial letters of a syllable such combinations as 'pr,' 'br,' 'kr,' but even between these a semivowel steals in. 'S' occurs only in the dog-call 'isū,'* 'h' only in one or two foreign words. Writing about Dippil, Dr. F. Müller says, "In the vocabulary of Rev. W. Ridley, there are indeed words in which the 'th' and 'dh' appear, but we believe the existence of these sounds in an Australian tongue doubtful and due to imperfect apprehension."† Dr. Müller's distrust is perfectly groundless. An English ear cannot be deceived in the sound of 'th,' it is a characteristic Australian sound, and in Kabi, of which Dippil is the nearest neighbour and almost the parallel, 'th' is pronounced exactly as in English *father*. The sound of 'dh' would be best illustrated by the value which would result from the 'th' in English *that* being preceded by a distinct 'd.' The Kabi 'v' is the equivalent of 'b' in some other dialects. Reduplication of consonants is frequent, each member of the pair being distinctly enunciated.

THE NOUN.

Number is denoted not by inflection, but by an adjective added. Gender is not marked by inflection excepting that there is a trace of ' -kan ' or ' -gan ' as a feminine termination in proper names and in the term 'nulangan,' *a mother-in-law,*‡ perhaps derived from 'yiran' or 'yirkan,' *a woman*. In all other instances such words as *man, woman, mother* are required to indicate the sex. Case is expressed by abundant terminations.

---

* The dog-call is "ai, ai, aiū, isū," aiū is a New Guinea word meaning *come*. The name of the dog is from New Guinea and no doubt the call was introduced with the animal.

† "Grundriss der Sprachwissenschaft," vol. ii. p. 42.

‡ 'Nulang,' *son-in-law*, 'nulanggan,' *mother-in-law*.

Probably the nouns are divisible into declensions distinguishable by the stem endings, but I am unable so to classify them. In nouns and pronouns the usual duplicate forms of the nominative occur, the one denoting the subject simply, the other the subject as active agent.

NOUN DECLENSION.

I employ the word 'ye̱ramin,' *horse*, because it is virtually a Kabi word, although applied to an imported animal, and because I am sure of important modifications to which it is subject. The terminations in this particular word about which I am uncertain, but which I have set down from analogy, are indicated by an asterisk, the analogies being supported by verified examples.

Nom.	*simple* . . .	ye̱ramin, *a horse*		dhakkẽ, *a stone*
„	*agent* . . .	ye̱ramin-dō		dhakke-rō
Gen.	. . . .	ye̱ramin-nō *		kung-u, *of the water*
Dat.	*to*	ye̱ramin-nō *		dhakkan-nō, *to the rainbow*
	to go for .	ye̱ramin-gō		
Acc.	. . . . .	ye̱ramin-na *		nguin-na, *the boy (object)*
Abl.	*because of*	ye̱ramin-ī		
„	*interest in,*	ye̱ramin-kari * or		wabun-gari, *on the stump*
„	*along with,*	gari		
„	*or upon*			
„	*instrument*	—		dhakke-rō, *with a stone*
				kuthar-ō, *with a club*

Other examples illustrating case are—

> *At or in*   no̱lla-nō, *in the waterhole*
> ngurun-nī, *by day*
> kira-mī, *at the fire*
> no̱lla-nī, *in the waterhole*
> kira-ba, *with or in the fire*
> nirim-ba, *in the middle*

According to Rev. W. Ridley the particle 'di' may be prefixed to indicate *of* or *from*, thus 'dhan di Boppil,' *a man of Boppil*.

PRONOUNS.

The pronoun is abundantly inflected and is of the common type in first and second persons singular and first and third plural. Gender receives no sound mark.

First Person.

	Singular.	*Plural.*
Nom. simple	. ngai	ngali or ngalin
„ *agent*	. ngadhu or adhu	ngalindō
Gen. of poss.	. ngañunggai	ngalinngūr or ngalinnō
Dat.	. ngaibọla	ngalingō
Acc. acted on	. nganna	ngalin

Second Person.

Nom. simple	. ngin	ngulam
„ *emphatic*	ngindai, nginbilin	
„ *agent*	. ngindū	
Gen.	. ngiñọnggai	ngulamō
Dat. motion to	nginbọla, nginbango	ngulambọla
Acc. acted upon	nginna	
also after give		ngupu, *you all*

Third Person.

Masculine, Feminine, and Neuter.

Nom. simple	. ngunda	dhinabu
„ *agent*	. ngundarō	dhinaburō
Gen.	. ngundanō	dhinabunō
Dat.	. ngundabọla	dhinabubọla, dhinabunga

Duals.

ngalinngin, *lit. we-thou,* used for *thou and I*
ngọlọm, *another and I*
bula, *you two*

There is no relative pronoun. For demonstrative the third personal is used, and also the words ' karinga,' *this one,* ' kọra-dhu' *that one.* To give a reflexive significance ' mitdhi,' *self,* follows the personal pronouns.

Interrogative Words.

Nom. simple	. ngangai, *who*
„ *agent*	. ngandō
Gen.	. ngañunggai
Dat.	. ngangaibọla, *to which place, whither*
Dat. and Acc. .	. ngangaimini, *whom* or *which*
Nom. simple	. miñanggai, *what*
„ *agent*	. ngandō, *what*

minani, *why*
minalū, *why*
miñama, miñamba, *how many*
miñamanō, *during*
minanggo, *how*
miñanggai, *what*

weño, *when* or *where*
weñamba, *whether or not*
weñobọla, *when, at what time*
weñomini, *where ever*
wandhurathin, *why*

'-amba' is a termination denoting *uncertainty, possibility,*
and is sometimes affixed to participles as well as to adverbs.

THE ADJECTIVE.

The language wants the article. The adjective is usually
undistinguishable by sound-sign from the noun, but a common
adjectival termination is '-ngur,' shortened sometimes *to* '-ngu.'
Adjectives can be formed from most nouns by affixing this post-
formative, the original meaning of which is not clear, but the
affix corresponds with the sign of the Kamilroi genitive, also
found in Kabi. Another adjectival ending in Kabi is '-dhau,'
by the addition of which certain nouns become adjectives. The
adjective is indeclinable. It is generally compared by the help
of such adverbs as 'karva,' *very.* Another mode of comparison
is to single out an individual and say of it *this (is the) large* or
this (is the) good, and so on according to the particular attri-
bute.

With the exception of the interrogatives enumerated already
and a few adverbs of place terminating in '-ni' and '-na' the
adverb has no phonic index. Those in '-ni' and '-na' may be
regarded also as locative cases of nouns. When a connective
is used, which is rarely, 'nga' answers for *and*, and if I mistake
not another mode of uniting ideas is to sustain considerably
longer than usual the final letter of a word.

NUMERALS.

The numeral system is binary. To express a number
higher than two the terms for one and two are combined as
may be necessary. 'Kalim' or 'kualim,' *one,* 'bulla,' *two* 'bulla
kalim,' *three,* 'bulla bulla' or 'bulla kira bulla,' *four.* The
enumeration may be conducted higher after the same manner,
but generally numbers above four are expressed by 'gurwinda'
or 'bonggan,' *many.*

THE VERB.

. The verb has various forms, as *Simple, Reciprocal, Causative,
Intensive.* But in certain instances what might be regarded as
a special form might equally be regarded as a distinct derivative

word from the simple form. Although regular examples may
be given there is a capricious irregularity about the moods and
tenses. Infinitive, Indicative, Suppositional, and Imperative
moods are distinguishable with well-marked terminations. The
infinitive and indicative may, however, be said to overlap. Tense
as indicated by termination is very wavering, the same forms
serving on occasions for present, past, and future time. There
is a clearly marked preterite, which is also a perfect participle,
terminating in '-n,' with 'a,' 'i,' or 'u' as preceding vowel.
The infinitive serves as imperfect participle, and there is also a
verbal noun. The shortest and simplest form is the imperative. Often it is one open syllable, it rarely exceeds two, but
sometimes adds '-morai.' Its termination is always in vowel
sounds. The general verbal notion is expressed by the infinitive
index, which is usually '-man,' '-mathī,' or '-thin.' Some verbs
may have an infinitive in two of these endings, thus there is
'yanman' and 'yanmathī,' *to go*, 'ñindaman' and ñindathin,' *to
enter*. The difference between the significance of '-man' and
'-mathi' is slight, if any, but as compared with '-thin,' the two
former indicate *state* or *inactivity*, the latter *action* or *motion*.
Person is not distinguished by sound, but has either to be
inferred or the pronoun is expressed and precedes the verb.
Conjugation is by means of prefixes, affixes, and what may be
called infixes. The prefixes generally are of adverbial force,
the affixes impart the modal, temporal, and participial significa-
tion, and the infixes may be regarded as possessing *formal*
power, expressing generally causative and intensive variations
of the sense, only it should be observed that the index of the
reciprocal *form* is terminal.

The following exemplify the use of prefixes—'biyaboman,' *to
come back*, from 'biya,' *back*, 'baman,' *to come;* 'yīkiyaman,' *to
answer*, from 'yīki,' *the same, likewise*, 'yaman,' *to speak ;* 'wuru-
boman,' *to come out*, 'wuru,' *out*, 'baman' *to come ;* 'yīvarī,' *to
put, to make*, is probably derived from 'barī,' *to bring*, and is
varied to 'mīvarī,' *to put away*, 'wuruyīvarithinī,' *to put out*.
In 'bīwathin,' *to play*, 'wathin' means *to laugh*, and 'bī-' is an
intensifying preformative, in 'bīyelī,' *to cooey*, 'yelī' means *to
shout*, 'bī-' has an intensifying or prolonging force. In 'bīdha-
līnda,' *to cause to drink*, the initial syllable transforms the Simple
into the Causative Form, or rather helps to do so, for '-lī' and

'-da' are also concerned in the change, 'dhathin' being the vocable meaning *to drink*.

The following are examples of affixes—'man,' '-mathī,' '-thin,' regular signs of infinitive, also of imperfect, indicative, and participle. '-an,' 'un,' '-in,' signs of preterite, perfect participle, and passive sense. '-ra,' '-thin,' '-thinī,' futurity and possibility. '-na,' '-nga,' '-ga,' 'da,' '-ngai,' marks of imperative mood. '-aio,' 'ati' distinguish the suppositional mood. '-na,' '-ba' are gerundive and participial (imperfect) signs. '-īra' has the sense of forcing or pressing. '-iū' implies irregular movement as exemplified in 'kauwaliu,' *to search*, 'maliu,' *to change*, 'yandiriu,' *to perambulate*. '-mathin,' '-bathin,' '-wathin,' transform other parts of speech into verbs and impart the significations respectively of (1) *purpose*, (2) *becoming*, (3) *holding* or *making*. '-yulaiyu' is the index of the Reciprocal Form, *e.g.*, 'baiyī,' *to strike* , 'baiyulaiyu,' *to fight*, *i.e.*, to *strike one another*.

Infixes.—Such terminations as 'man,' 'mathī,' &c., express the general verbal sense, having some such force as *do* or *make*. Without removing this general verbal sign one or more syllables may be interposed between it and the stem; this is the usual mode of indicating the Causative and Intensive Forms. The word 'karī' means *here* or *in ;* 'karīthin' is *to enter*, with preterite 'karīn.' The termination '-thin' differs little from '-man' or '-mathī' in force; there is also a verb 'kari-na-man' and another 'karin-di-mi,' both meaning *to put in*, '-na' and 'di' are the Causative indices. The word 'buwandiman' means *to herd*, lit. *to cause to stop;* it is thus compounded, 'buwan,' *to stand*, 'di,' causative particle, '-man,' verbal sign. The infix '-li' is introduced to imply *doing well, progress, advantage*. Examples, 'yangga,' *to make*, 'yanggalīnoman,' *to allow*, from 'yangga,' '-li,' *to advantage*, '-no,' *permission*, 'man' verbal sign. 'Womba' means *to lift*, 'wombalīman,' *to fall upon*, 'wombalīn,' *carrying*, the word 'wombalīmaraiō' may therefore be thus analysed, 'womba,' *to lift*, '-li,' *motion*, '-mara,' sign of futurity, '-aio,' mark indicating supposition.

One kind of modification yet remains to be noticed—viz., reduplication. This is the usual sign of the Intensive Form, *e.g.*, 'yelīman,' *to shout*, 'yelelīman,' *to speak quickly*, 'dhoman,' *to eat*, 'dhandhoman,' *to gnaw*, 'dhomma,' means *to catch*, 'dhomathin,' *to hold, to grip,* 'dhommoman,' *to marry, i.e., to catch and hold fast!*

Mr. Threlkeld in his "Key to the Structure of the Aboriginal Language" is overpowered and carried away by a mystic propensity when he affirms that to the aboriginal mind particular letters or syllables have a sense inherent in the sound. However, from examples which he gives, as well as from the above, it is evident that a letter or syllable may be elegantly introduced to shade delicately the meaning of the verb. But such letters or particles are no doubt remnants of words too much broken down to stand alone.

PARADIGMS.

Forms.

Simple.	*Causative.*	*Intensive.*	*Reciprocal.*
yeliman, *to call*		yeleliman, *to*	
yaman, *to speak*		*speak quickly*	yathulaiyu, *to*
buwan, *to stand*	buwandiman		*converse*

Moods.

Inf. and Indic.	*Imperative.*	*Suppositional or Subjunctive.*	*Verbal Noun.*
yeliman, *to speak*	yeli		yelinba
baman, *to come*	ba		
buwan, *to stand*	bubai	boiu	

Preterite, Perfect Participle, and Passive, 'ya'an,' *spoken,* 'ban,' *come.*

Auxiliary verbs are unknown; temporal differences are generally expressed by an adverb of time.

This may be the best place to show the relation which Kabi bears to the other dialects of the N. S. Wales and S. Queensland class, chiefly to Kamilroi and Wiradhuri. The very name Kabi is the local equivalent of Kamil; the Kabi people would drop the final 'l,' as in the word 'mi,' *eye,* of which the Kamilroi form is 'mil.' From the sea coast at Maryborough for about 450 miles inland, in a south-westerly direction, the natives may be regarded as virtually one tribe lingually. The following are particular analogies:

KAMILROI.	KABI.
andi, *who*	ngandu, *who*
minya, *what*	minyanggai, *what*
gir, *verily*	givir, *verily*
yeül, *merely*	yul, *in vain, gratuitously*

KAMILROI.	KABI.
guru, *round*	kuri, *round*
baoa, *back*	biya, *back*
taon, *earth*	dha, *earth*
tulu, *tree*	dhu, *tree*
moron, *alive*	murrumurru, *full of life*
giwir, *man*	kivar, *man*

Many other examples might be adduced.

I shall conclude this sketch with a table of case-endings; for those of the first four dialects I am indebted to Dr. F. Müller's "Grundriss der Sprachwissenschaft," the fifth I add to show the place of Kabi in the group.

	LAKE MACQUARIE.	WIRADHURI.	KAMILROI.	TURRUBUL.	KABI.
Nom. agent	-to	-tu	-du	-du	-do
Dative	-ko	-gu	-go	-ngu	-no and -go
Genitive	-ko-ba	-gu-ba	-ngu	-nu-ba	-no and -ngu
Ablative	-tin	-di	-di	-ti	-ni
Locative	-ta-ba	—	-da	-da	-ba

The verbal definitive elements differ considerably. A comparison of interrogative words leads to the same conclusion—viz., that all the members of this group are very closely related.

SPECIMEN OF KABI WITH TRANSLATION.

DHAK'KE' KUNDA'NGUR.

Pebbles (of) Koondangoor (a place).

Dhan ngain dhak'keni nol'lani nyenaman.
(*To the*) *blackfellow always pebbles* (*in his*) *inside are.*
Piri ngim búyū kain gillin.
(*In the*) *hand bones calves head nails.*
Ngin kárunda yin'malo ngin'bango dhúngun kar'ithin.
Thou floating (?) *remain* (*to*) *thee* (*in the*) *stomach* (*they*) *enter.*
Ngin wa bai'yiro yūn'maman ngin man'ngūrbathin.
Thou not aching (*wilt*) *lie thou* (*wilt*) *become-full-of-vitality.*

NGAN'PAI BAT'YIMAN.

Pebble Finding.

Ngin yun'mai dhu'mo tar'vano. Ngin'du kui'bi vrg'nga
Thou lie down (*a*) *tree under. Thou* (*a*) *whistling wilt hear*
ngan'pai ngin'bola nyin'dathin. Ngūn'da dhilīl'bangur nyin'daman.
the pebble (*to*) *thee* (*shall*) *go in. It noisily* (*shall*) *go in.*
Ngin man'ngur nye'naman wa ba'luman.
Thou full of vitality (*wilt*) *be not* (*wilt*) *die.*

BAI'YI YANGGAL'ITHIN.

Pain Curing.

Ngai   we'nyo   bai'yīngūr   mu'ru mu'ru   ngan'na   būn'bithin
I   *if (or when)*   *sick*   *(the man) full-of-life*   *me*   *(will) suck*
              dhak'ke   ngan'na   būn'mathin.
              (the) pebble   *(from) me*   *take out.*

DHAK'KAN MAN'NGUENGUR.

(The) Rainbow capable of imparting vitality.

Ngin   bon'na   bai'yīngur   yen'na   yūnmathin   kūngu   karáno.
Thou   *when*   *sick*   *go*   *lie down*   *(at the) water's*   *edge.*
              Ngin   bai'yī-yang'galīthin.
              Thou   *wilt be cured.*
       Dhan   dhak'kanno   ngan'pai   wom'ngan
(The black) man   *(to the) rainbow*   *pebble*   *gives*
         dhak'kan   dhan'no   bū'kūr   wu.
         the rainbow   *(to the black) man*   *rope*   *gives.*

DHAK'KAN.

(The) Rainbow.

Dhak'kan   wa'rang   ngun'da   kor'aman   ngu'in
(The) rainbow   *(is) wicked*   *he*   *stole*   *(a) boy*
         dhī'kui,   kar'vana   wom'ngan   mūl'lū.
         half-caste,   *another*   *gave*   *black.*
Ngun'daro   kom'ngan   ngū'ina   tūn'bano   nol'lano
He   *took*   *(the) boy*   *(to the) mountain*   *(a water)-hole*
   karin'dimi .   Nollani   ngū'in   nye'naman ;
   (he) put (him) in.   *(In the) hole*   *(the) boy*   *is ;*
              ngu'rūni   wū'rūboman.
              during the day   *he comes out.*

THE LANGUAGE OF WESTERN AUSTRALIA.

Authorities : Captain (now Sir George) Grey* and Mr. G. F. Moore.†

This language is spoken in the neighbourhood of Perth, and
with slight diversity in the greater part of the south-west of
Western Australia.   So far as appears it is the most rudimentary
and analytic of Australian languages.

PHONIC ELEMENTS.

Vowels.

a   ā   ǎ

e   o   ō   ǫ (as in *ton*)

i   ī   u   ū

* " Vocabulary of the Dialects of South West Australia."
† " Ten Years in Western Australia."

Diphthongs.
ai au oi ua uo

Consonants.
k g   ng
t d tch or tz, dj   y r l n ñ
p b v (in one word only)   w m

There is an aversion to 'r' and 'l' at the beginning of words.
The distinction between surd and sonant letters is undecided.
The noun seems destitute of case-endings. The names of social
relations have a plural form in ' -mun' if the singular end in a
vowel, in ' -gurra' if the singular end in a consonant; 'mun' is
an abbreviation of 'munda,' *altogether, collectively,* ' -gurra' is
probably derived from 'garro,' *again.* 'Migalya' is the plural
of 'migal,' *a tear.* The comparative of adjectives is formed by
reduplication, the superlative by the addition of ' -jil' or ' buk.'
The pronouns, besides having three forms of dual for the three
persons, have also a trial number for the first person.
Possessive pronouns are formed from the personal by affixing
' -uk,' excepting in the second person singular. This -uk as a
sign of possession unites the eastern and western languages.
This affix effects the same result in compound expressions, where,
however, it sometimes changes to ' -ung.'

The verb is exceedingly simple. The preterite is formed by
adding ' -ga,' the participle present by affixing ' -een' or '-ween'
to the present tense with the occasional interposition of a vowel
at the junction thus :

Present indicative .   . yugow, *stand*
Preterite   .   .   . yugaga
Present participle .   . yugoween

The preterite has three forms relating respectively to the immediate past, the sometime past, and the remote past. These are
distinguished by *prefixing* to the regular preterite the particles
'gori,' 'garum,' 'gora,' respectively. There are two futures, a
near and a distant, distinguished by the words 'boorda,' *presently,*
and ' mela,' *in the future,* which follow generally the infinitive
mood, occasionally the present participle, but are not incorporated with the verb. The word 'ordak,' signifying *to intend,*
is also affixed to verbs to denote that the action is purposed.
There is likewise a past participle which is not specified. There

N

is no phonic mark of number in the verb. The different persons
are indicated by employing the pronouns.

This language favours. the combining of words to an almost
indefinite extent. The word commonly employed to give unity
to compounds is 'midde,' *the agent* or *agency*, and all verbs may
be rendered substantive by the addition of this word. For
example, 'yungar barrang midde' is the *horse,* or literally the
people-carrying agent, 'mungyt barrang midde,' the 'mungyt'-
getting-agent or *stick for hooking down the Banksia cones.*

There are combinations observable in the verb which seem
elementary forms of the more complicated structure in the east
of the continent, thus :

> yugow-murrijo (literally *to be, to go*), *to run*
> yugow-murrijobin, *to run quickly*
> yulman means *in turn, in return*
> wangow, *to speak;* yulman-wangow, *to answer*
> yonga means *to give;* yulman-yonga, *to exchange*

'yulman' is singularly like the reciprocal sign in the eastern
dialects, which in Kabi for instance is ' yulaiyu,' but in the east
it is affixed to the verbal stem.

PRONOUNS.

First person.

	Singular.	*Plural.*
Nom. simple .	nganya	ngannil
,, *agent* .	adjo or nadju (ngadju ?) ngadjul	
Gen. . . .	nganaluk, *also* nganna	nganiluk
Acc. . . .	nganya	ngannil

Captain Grey gave to nadju the sense of *I will*, but probably
as in other cases it expresses the agent; a similar remark applies
to the corresponding form in second person.

Second Person.

	Singular.	*Plural.*
Nom. simple .	nginni	nurang
,, *agent* .	ñundu or ñundul	
Gen. . . .	nunoluk	nuranguk, ngunullang, ngunaluk
Acc. . . .	nginni	

Third person,

	Singular.	*Plural.*
Nom. . . .	bal, *he, she, it*	balgun, bullalel
Gen. . . .	baluk, *her;* buggalong, *his*	balgunuk
Acc.		balgup
Dat. . . .	buggalo, *to him,* ballal, *he himself*	

DUALS.

Brother and sister, &c.	Parent and child, &c.	Husband and wife, &c.
1st person . ngalli	ngala	nganitch
2nd person . ñubal	ñubal	ñabin
3rd person . boola	boolala	boolano

nganuama, *we two* (*brothers-in-law*)

TRIAL *1st person*, ngalata, *we three*

There are only three numerals, 'gain' or 'kain,' *one*, 'gudjal,' *two*, 'ngarril,' 'warring,' 'mow,' 'murdain,' *three*. Higher numbers are expressed by 'warring,' *a few*, or 'boola,' *many*. 'Boola' is evidently the same as the eastern term for *two*, as it is used for a dual pronoun.

Interrogatives—Pronouns.

Nom. simple . ngauni, *who*		nait, *what*
„ *agent* . ngando, ngandul, nginde		yan, „
Gen. . . . . ngannong, *whose*		

Adverbs.

winji winjala (windyi, windyala), *where*, naitjak, *wherever*.

THE DIYERI LANGUAGE.

Authority: Mr. Samuel Gason's "The Dieyerie Tribe of Australian Aborigines."

The Diyeri language is spoken between Cooper's Creek and the north-east shore of Lake Torrens, in South Australia, but not far from the Queensland and New South Wales boundaries. Mr. Gason's vocabulary does not supply much data for arriving at the structure of sentences, the examples of syntax being unfortunately too meagre to admit of our deducing noun declension from them. The verb seems to be conjugated very simply and with a suspicious regularity. The language is of a very elementary, compounding character, and in this respect stands midway between the languages of the extreme west and east respectively, being more closely related to the latter. The personal pronouns and some of the interrogative words unite both extremes with the mean. The reciprocal sign of the west, 'yulman,' is well worth comparing with 'mullauna,' *one another*, of the Diyeri, '-ulunni' of Kamilroi, 'yulaiyu' of Kabi, '-lan' of Lake Macquarie, and 'lana' of Wiradhuri, all reciprocal verbal signs; the Kabi and West Australian forms seem to give the original type as something like 'yulain,' which may be com-

pounded of two pronouns, 'ngali-ngin,' *we-thee,* or the like.
Diyeri is rich in determinant elements, easily recognisable and
separable, and usually, but not invariably, post-formative.

PHONIC ELEMENTS.

Vowels.

a ă
  e o ǫ (as in English *ton*)
i ī   ū

Diphthongs.
ai au

Consonants.

k g h   ng
  t d   l r y n
th ch
p b   w m

The words terminate with *vowel sounds only,* they begin with
vowels or consonants, but the latter must be single. Such combinations occur internally as rd, rt, rk, kr, dr, ldr, ndr. Diyeri
therefore agrees fairly in phonesis with the eastern languages
generally, but is even smoother and more vocalic.

PRONOUNS—*First Person.*

	Singular.	*Plural.*
Nom. (*agent?*)	athu	ali, yana, uldra
Gen.	ani	yanani, uldrani
Gen. or dat.	akunga	
Acc.	ani, ni	ali

Second Person.

	Singular.	*Dual.*	*Plural.*
Nom.	yondru	yula	yura, yini
Gen.			yinkani
Acc.	ninna		

Third Person.

	Singular. Mas.	*Feminine.*	*Neut.*	*Dual.*	*Plural.*
Nom.	noalia	nandroya	ninna,	bulyia	thana
Gen.	nunkani	nankani			thanani
Gen. or dat.					wurra, wirri, yinkani
Acc.	nulu	nania			thaniya

'ninna' is also demonstrative, *this.*

There are definitive elements affixed to substantives to
signify *my,* as 'api-ni,' *my father,* '-ni' is a general genitive or

possessive termination with '-li' as probably an alternative form.

Possessive forms are evidently declined, *e.g.*, 'yinkari,' *yours,* 'yinkani-gu,' *of or to yours.*

It is much to be regretted that data are lacking from which the declension of substantives might be tabulated. The noun is probably rich in cases, as may be inferred from such compounds as the following, 'bumpu-nundra,' *almost a blow,* 'nundra,' *a blow,* 'bumpu,' *almost;* 'moa-pina,' *very hungry,* 'moa,' *hunger,* 'pina,' *great;* 'kurnaundra,' *relating to a blackfellow,* 'kurna,' *blackfellow,* 'undra,' *relating to.* A genitive is observable in '-lu,' *e.g.,* 'pinya,' *armed party,* 'pinyalu,' *of the armed party.*

Interrogative Words.

Nom.	. warana, *who*	
Gen.	. wurni, wurniundru, *whose*	
Acc.	. wurungu, *whom*	
	whi, *what*	
	wodau, *what, how*	
	wodani, *what is it like?*	
	wodaru, *what do you say?*	
	wodau, *how*	mina, *what*
	wodaunchu, *how many*	minani, *what else*
	wintha, *when*	minandru, *why*
	winthuri, *whence*	minarnani, *why*
	wodari, *where*	

Adjectives do not seem to be distinguishable by any vocal sign, but comparison is marked by added definitive elements, thus 'wurdu,' *short,* 'murla,' *more* 'muthu,' *most,* 'wurdu-murla,' *shorter,* 'wurdu-muthu,' *shortest.*

NOTATION.

'Curnu,' *one,* 'mundru,' *two,* 'paracula,' *three.* The numeral system is virtually binary. Twenty is expressed by 'murrathidna,' *hands-feet,* for any number over twenty an indefinite word signifying multitude is employed.

THE VERB.

The structure of the verb so far as we can judge is exceedingly simple. To indicate the person the pronoun is prefixed unabridged. There are simple and reciprocal forms, the latter having the termination 'mullāna.' The simple form has infinitive, indicative, and imperative moods, and participles perfect and imperfect. The following is the conjugation of the verb 'yathami,' *to speak*, parallel with which I place the Kabi verb 'yamathi,' also meaning to speak, in order to show the singular likeness and close relationship of the word and its modifications :—

DIYERI.	KABI.
yathami, *to speak*	yamathi, *to speak*
yathunaori, *has spoken*	yamarandh, *spoken*
yathi, *have spoken*	ya'an, *spoken*
yathunawonthi, *had spoken*	wonai yamathi, *have done with speaking !*
yathulāni, *will speak*	yathin, *will speak*
yathala, *speak*	ya, *speak*
yathamarau, *speak* (imperatively)	ya, yamorai (by analogy of other Kabi verbs), *speak* (imperatively)
yathuna, *speaking*	yathinba (by analogy as above) *speaking*
yathamullāna, *quarrelling together*	yathulaiyu, *conversing*

The stem radical of the above verb is evidently 'yath,' the original infinitive form containing the theme or notion of the action was evidently 'yathamathi,' the medial 'a' being introduced as a connective. This leads to the conclusion that '-mi' or '-mathi' is a verbal definitive which was probably once a verb meaning *to do* or *make*, like the '-ed' of the past tense in English regular verbs, which is *did* phonetically decayed. Another very suggestive comparison may be made between a Diyeri verb and its analogues in Kamilroi and Kabi :—

DIYERI.	KAMILROI.	KABI.
wima, *put*	wimi, *put down*	womngan, *give*
		womngathi, *give*
wimuna, *putting in*		wiyin, *given*
wimarau, *put in* (imperative)	wimulla, *put down* (imperative)	womorai, *give* (imperative)
yinkumullāna, *giving each other*	wiulunni, *to barter*	wiyulaiya, *to exchange*

In Diyeri 'wima' has no reciprocal, I therefore show the reciprocal of 'yinkuna,' *giving*. The original infinitive of the verb to give is probably 'wīyimathi' or 'wīyingamathi,' 'wī' or 'wīyi' being the stem. But what is specially noticeable is the close agreement of the imperative forms. The Kabi imperative is generally the simplest and shortest form of the verb, but it has also a form in '-morai' as here represented, which appears to be emphatic, and the force of '-morai,' as also of the terminations in the other dialects, '-marau,' '-mulla,' is evidently *do*. In my contribution on the Kabi in Mr. Curr's work this passage occurs, "The ending '-morai' appears in some imperatives given in the table of conjugations. As we also find an infinitive termination '-moraman,' it seems to me that '-mor' was the stem of a verb now obsolete which was almost equivalent to the verb *do*, and it now exists merely as an intensifying ending."[*] I was not then aware that 'ma' or 'mara' was a verb in Wiradhuri meaning *to do* or *make*, but is it not highly probable that parts of that verb have become the regular terminal marks in different parts of the verb in many dialects, as, for example, '-ma,' '-mi,' 'mathi,' '-man,' indices of the infinitive, and '-morai,' '-marau,' &c., of the imperative, and further is it not also probable that these terminations are radically connected with the Malay 'men' prefixed to words to transform them into verbs?

LANGUAGE AT MACDONNELL RANGES.

Authority: Rev. H. Kempe, by kind permission, "Transactions Roy. Soc. S. Australia," 1890–91.

With but slight variations this language is spoken from the Finke River eastward to Alice Springs and extends south to the Peake. It is the central language of Australia. Some of its most striking features are found in the dialects near the S.E. corner of the Gulf of Carpentaria on the Norman and Palmer Rivers, notably a preference for initial vowels and certain vocables uniting these languages and distinguishing them from others.

PHONIC ELEMENTS.

Vowels.

a o

e ū

ĭ u

Diphthongs.

au ai oi

---

[*] Curr, "The Australian Race," vol. iii. p. 189.

Consonants.

k g h ng
t d r l y n ñ
p b w m

Any of the letters may be initial. Vowels are preferred, but in many cases they appear to have become initial by the elision of a consonant, especially 'k.' The terminal letter—except in the vocative of nouns and the imperative of verbs—is always 'a,' in which respect the dialect is singular, the nearest approach to it being the dialect of East Tasmania.

THE NOUN.

The usual three numbers obtain, singular, dual, and plural. The dual is formed by adding '-ntatera.' With terms for persons another form is used in '-nanga,' *e.g.*, 'wora,' *boy*, 'worananga,' *the two boys*. The plural is formed by adding '-irbera' or '-antirbera' to the singular

There are six cases. When related to an intransitive verb the nominative is unchanged; when related to a transitive verb it takes the termination '-la,' *e.g.*, 'wora-la gama,' *the boy gets*. The genitive is formed by adding '-ka,' as 'kata-ka,' *of the father*. The dative ends in '-na,' the ablative in '-nga.' For the accusative there is no change. The vocative is in '-ai.'

Derivative substantives are formed by adding '-ringa,' lit., *to be at home at*, and by reduplication from verbs.

There is no article.

THE PRONOUN.

All the possessive pronouns are perfectly declined like the nouns.

The possessive pronoun, first person, is thus declined:

Nom.	.	. nuka
Gen.	.	. nukanaka
Dat.	.	. nukana
Acc.	.	. nuka
Abl.	.	. nukananga

The dual of the third person possessive is 'ekuratera,' *their two*. All persons of the plural are declined like the singular.

Of the personal pronouns the forms 'ata' or 'ta,' 'yinga' (first person), 'unta' (second person), are used only in the nominative case. The third personal are regularly declined in singular, dual, and plural numbers.

'Nukara,' *myself*, and 'etnikara,' *one another*, are either reflexive or reciprocal as the verb may determine.

The demonstratives are—

nana, *this*	tana, *that*
nanatera, *these two*	tanatera, *those two*
nanirbera, *these*	tanirbera, *those*
nanankana, *these*	tanankana, *those*

'Nakuna' and 'arina' are also used for *that*.

The demonstrative pronouns are declined like the possessive.

The interrogative pronouns are 'nguna,' *how*, and 'iwuna,' *what;* the dual and plural are formed as in demonstratives, but when declined the inflections are medial.

As a substitute for the relative pronoun, which, as usual in Australian languages, does not exist, the demonstrative pronoun is repeated or else the relation is implied in the participle.

INDEFINITE PRONOUNS.

arbuna and tueda, *another*	nkarba, *a few others*
arbunatera, *two others*	tueda ka tueda, *others*
arbunirbera, arbunankana, *others*	ninta mininta, *one by one*
urbujarbuna, *some others*	

The above are regularly declined. 'Arbuna' is specially interesting, as suggesting the original significance of the commonest Australian term for the numeral one. 'Arbuna' is the analogue of 'karva' and 'karvano' (*another*) in the Kabi dialect of South Queensland.

The following are co-relatives used only in nominative cases.

ntakina, *how, in what way*	lakina, *thus, in this way*
ntakinya, *how many*	lakinya, *so many*
ntakata, *how big*	nakakata, *so big*

ADVERBS.

Derived adverbs are formed by adding '-la.' 'Nana' is *here*, 'avina,' *there*. Numeral adverbs are 'arnkula,' *the first;* 'ninta ranga' and 'ninta ngara,' *once;* 'tera ranga,' *twice;* 'urbuja ranga,' *sometimes.*

The majority of the conjunctions are combined with the verb.

THE VERB.

There are three tenses, the present, marked by '-ma' affixed to the stem, which is also the infinitive ending. The perfect is formed by adding '-ka' or '-kala' to the root and sometimes '-jita'; the future adds '-jina.'

VOICE.

The middle voice takes its sign, '-la' or '-li,' in the middle of the verb—*e.g.*, 'ta nukara tu*la*ma,' *I beat myself.* The reciprocal form terminates in '-rama' for dual and '-rirama' for plural.

There is no proper passive. Passivity is expressed by employing the subject with the active form and having the suffering object in the dative—*e.g.*,

atula	worana	tukala
by the man	*to the boy*	*is beaten*

NUMBER

There are three numbers :—singular, dual, and plural. If no pronoun be expressed they can be distinguished by termination of the verb. Person is not noted. In intransitive verbs the dual termination is '-rama,' the plural '-rirama.' With transitive verbs the dual and plural are formed by using 'nama' (to be) and 'lama' (to go) as auxiliaries.

In the middle voice the dual and plural double the particle '-la'—*e.g.*, 'ilinakara tu*lala* narama,' *we two beat ourselves.*

MOOD.

There are three moods, indicative, conditional, and imperative. The conditional is formed by adding '-mara' to the stem, as 'ta ilkumara,' *I should eat.* The imperative is formed by adding '-ai' to the stem, as 'tu-ai,' *beat;* 'ilgula nar-ai,' *you two eat;* 'ilgula narir-ai,' *eat ye.* Another form, signifying to do the action quickly, is composed by inserting the syllable '-lba' between a duplication of the root—*e.g.*, 'tu-lba-tu-ai, *beat quickly.* The moods have usually a negative as well as a positive form. Thus :

Positive.	*Negative.*
ta tuma, *I beat*	ta tuyikana, *I beat not*
,, nyuka, *I drank*	,, nyuyinakana, *I have not drunk*
,, gayina, *I shall get*	,, gayigunia, *I shall not get*

PARTICIPLE.

The imperfect is formed by adding '-manga' to the stem, the perfect by adding '-mala,' the future by adding '-yinanga; —*e.g.*, 'tumanga,' *while beating;* 'tumala,' *after beating;* 'tuyinanga,' *shall be beating.*

Certain verbs such as 'nama,' *to be*, and 'lama,' *to do*, are used by way of auxiliaries. Their use is (1) to change transitive into intransitive verbs, (2) to form verbs from substantives and adjectives.

To express such modifications as are usually expressed by adverbs in European languages certain vocables are combined with the verb. For example, 'tula' is combined with 'wuma,' *to hear*, 'nama,' *to be*, 'lama,' *to go*, 'albuma,' *to return*, and so forth, as :

> tula nama, *to beat for a time constantly*
> tula lama, *I go to beat, signifies an action going on*
> tula tula lama, *to beat sometimes, quickly or hastily*
> tula uma, lit., *I hear to beat, means I beat once*
> tula albuma, *is arrived at another place beating*

Certain forms combine with the supine, as :

tuyikalama, compounded of supine 'tuyika' and 'lama,' *to go*, lit., *I go to beat*, used for *I beat downwards*; 'tuyinyama,' *I beat upwards*.

A figurative use of the forms occurs in the modifications of the verb 'ilkuma,' *to eat*, as :

> ilkuyikalama, *to eat in the evening*
> ilkuyinyama, *to eat in the morning*
> tuyikamaniyikana, *I beat again*
> tuyikerama, *going to beat*
> tuyilbitnima, *come to beat*
> tuyalbuma, *return to beat*
> tuyigunala, *to beat by-and-by*

From tutua, meaning *I desire to beat*, are derived such forms as :

> tu (I) tuatna lama, *I beat arrived at another place*
> ta tualbanama, *I beat sometimes arrived at another place*
> ta tualbuntama, *I beat running away*
> ta tuatalalbum, *returning homewards I beat on the road*
> ta tuatnilbitnilalbuma, *returning come near my home I beat*

Moods and forms connected with moods already given :

> ta tumalamakana namara, *I should not have beaten.*
> ta tumaranga *or* tumalanga, *I should have beaten.*

The reduplications or augmentations of the verb:

> tuyinabuta tuyinabuta, *I should beat again*
> tulinya tulindama, *to beat always*
> tulinya mbura   „   „   „
> tulatulauma, *to beat seldom*
> tuma, *I have finished beating*

By the simple verb 300 different phrases can be used; by modifications of the verb, these can be increased to 9000. By further changes confined to certain moods and tenses an additional 600 verbal phases are obtained, so that 9600 expressions may be derived from one verb.

FOREWORD TO COMPARATIVE TABLE

In the following table fifty-two lists of words are given. Of these, forty-two are Australian, three New Hebridean, two Torres Strait, and five Tasmanian. The aggregate number of English words is two hundred and twenty-five. The New Hebridean and a number of the Australian lists are fairly complete. One object of the table is to exhibit the relation subsisting among Australian dialects themselves, and their connection with the languages to the south, east, and north-east of Australia. The Australian dialects are grouped and, on the whole, graduated according to strongly marked resemblances. They are gathered towards the north-east, as the fingers of the hand are gathered towards the wrist.

AUTHORITIES

TASMANIA.—Vocabularies collected in Mr. E. M. Curr's "The Australian Race" and one from "Voyage de l'Astrolabe." The capitals in brackets indicate the following names:—D., Dove; E., Entrecasteaux; J., Jorgenson; L., Lhotksy; N., Norman; P., Peron; R., Roberts; S., Scott.

When not thus distinguished, the authority for the north dialect is "Voyage de l'Astrolabe," and for the others Dr. Milligan's lists.

AUSTRALIAN, VICTORIAN REGION.—Yarra River, Lal Lal, Ercildoune, Avoca River, Broken River, Gunbower, Warrnambool, were taken down by the writer (Rev. J. Mathew) from the lips of natives.

Mortlake, supplied by Miss Hood, Merrang, Hexham, Victoria (now deceased).

Booandik, South Australia, compiled from "The Booandik Tribe of South Australian Aborigines," by Mrs. James Smith.

Lower Lachlan and Murrumbidgee, contributed by Mr. Humphrey Davy, Glen Dee, Balranald, New South Wales.

Gippsland, taken down by Rev. J. Mathew.

Barwidgee, Upper Murray, contributed by Mr. John F. H. Mitchell, Khancoban, Corryong.

NEW SOUTH WALES AND SOUTH QUEENSLAND.—Woorajery Tribe, con
tributed by Mr. James Mitchell, Table Top, Albury, New South Wales.
    Wiraidhuri, Günther in Fraser's "An Australian Language."
    Port Jackson, the vocabularies of Captain Hunter's "Historical Journal
of the Transactions at Port Jackson," and Lieut.-Colonel Collins "New
South Wales."
    Awabakal, Threlkeld's "Australian Grammar."
    Kamilaroi, Rev. W. Ridley's "Kamilroi and other Australian Lan-
guages."
    Kabi, Rev. J. Mathew and Mr. W. Hopkins.
    Warrego River, contributed by Mr. W. Shearer, Brewarrina, New South
Wales.

WEST AUSTRALIA AND WEST CENTRAL.—Toodyay, Newcastle, West
Australia, contributed by Mrs. George Whitfield.
    Pidong   } "Journal of the Elder Scientific Exploring Expedition,
    Minning }   1891-2."
    Lake Amadeus, Mr. W. H. Willshire's "Aborigines of Central Australia."

SOUTH OF SOUTH AUSTRALIA AND EAST CENTRAL.—Narrinyeri, Rev.
George Taplin's "Folklore," and Mr. E. M. Curr's "The Australian
Race."
    Parnkalla, Schürmann's "Vocabulary of the Parnkalla Language."
    Adelaide, Teichelmann and Schürmann's "Grammar Vocabulary, &c.,"
and Dr. Wyatt from J. D. Wood's "Native Tribes of South Australia."
    Darling, Mr. E. M. Curr's "Australian Race."
    Diyeri, Mr. S. Gason's "The Dieyerie Tribe."
    Murunuda, South Gregory, Mr. Duncan M. Campbell, Glengyle, More-
land, Melbourne.
    Mythergoody, Cloncurry, Mr. W. G. Marshall, Fort Constantine, Queens-
land.

NORTH AUSTRALIA AND CENTRAL AUSTRALIA.—Larrikeya, Member of the
Roman Catholic Mission at Daly River, per Mr. George McKeddie, with
some words from T. A. Parkhouse's "Transactions of Royal Society of
South Australia," vol. xix.
    Woolna, Mr. E. M. Curr's "Australian Race," and Mr. T. A. Parkhouse.
    Daktyerat, Member of the Roman Catholic Mission, Daly River, per
Mr. George McKeddie.
    Kimberley, contributed by Mr. Joseph Bradshaw, Melbourne.
    Napier Range, contributed by Mr. W. W. Froggatt, Sydney, New South
Wales.
    Sunday Island, contributed by Mr. Joseph Bradshaw, Melbourne.
    Macdonnell Ranges, Rev. H. Kempe, Mission Station, Finke River, in
"Transactions of Royal Society, South Australia, 1890-91."
    Walsh River, Rev. J. Mathew.
    Bloomfield Valley, contributed by Mr. Robert Hislop, Wyalla, Bloom-
field River.
    Palmer River, Mr. E. M. Curr's "The Australian Race."
    Coen River } Contributed by Revs. J. Ward (now deceased) and
    Mapoon   } ·   N. Hey.
    Cape York, McGillivray's "Voyage of H.M.S. *Rattlesnake*."

Torres Strait.—Kowrarega, Prince of Wales Island, McGillivray's "Voyage
of H.M.S. *Rattlesnake.*"
    Saibai Island, Sir W. MacGregor's "Reports," specially forwarded to
the writer.

New Hebrides.—Aulua, Malikula, Rev. T. Watt Leggatt, Amy Gertrude
Russell Mission House.
    Nguna, Rev. Peter Milne.
    Aniwa, Rev. J. G. Paton, D.D., Melbourne.

With the view of securing consistency of spelling, I suggested
to my correspondents the following values of letters : The
consonants may have the same value as in English, only that *g*
should always be hard, as in 'go' or 'get.'   *Ch, j,* or *s* should
not be introduced unless absolutely necessary; *k* or *s* should
take the place of *c*; and neither *q* nor *x* should be required,
k answering for *q* and *ks* for *x*.   The initial nasal sound should
be expressed by *ng*.   *Dh* represents *th* in 'the.'
    The vowel sounds are—

a as in 'father,' 'man.'	*e* as in 'they' or 'net.'
ai like *i* in 'mine.'	*u* as in 'rude' or *oo* in 'food.'
i like *i* in 'pit' or 'ravine.'	*au* is used for the sound of *ow* in 'cow.'"

There has been a tendency, however, to independence in orthography.   The above represents with fair accuracy the values in
the lists which I have obtained myself, and will serve for general
guidance in reading the table.

Groups	Dialects	Sun	Moon	Star
Tasmania	East	pugganoobranah	wiggetena	teahbrana
	South	pallanubranah	weeta	romtenah
	West and N. W.	panubrynah	weenah leah	rhomdunna
	North	tegoura (?)	tegoura	moordunna (J.)
	Miscellaneous .	loina (J.)	luina (J.)	daledine (R.)
Victorian Region	Yarra R. . .	nyawain	mirnian	turtbairom
	Lal Lal . . .	mirri	yen	turtparom
	Ercildoune . .	nauwi	yen	dut bun-nauwi
	Avoca R. . .	nauwi	yen	turt
	Broken R. . .	ngamaik	mirnan	durt
	Gunbower . .	ñawi	wainwil	toort
	Warrnambool .	nganong	yathyar	ka ka dhirn
	Mortlake . . .	dhearn	koondarook	—
	Booandik, S. A.	karo	toongoom	tumanbangalum (*pl.*)
	Lower Lachlan and Murrumbidgee	nangaye, nung	wangupie	toorty, tingie
	Gippsland . .	wurîn	noran	brîl
	Barwidgee, Upper Murray	noweyu	bararoo	jeembo
N.S. Wales and S. Queensland	Woorajery Tribe, Upper Murray, N.S.W.	yerai	kupador	—
	Wiraidhuri .	ire	giwang	buddu (*pl.*)
	Turuwul, Port Jackson	coing (H.), wirri (R.)	yennadah (C.)	birrong (C.)
	Awabakal . .	punnul	—	—
	Kamilroi . .	yarai	gille	mirri (*pl.*)
	Kabi, Mary R.Q.	nguruindh, tirum	bapun	miringam, kalbar
	Warrego R., Q.	durrey	gheërn	myrring
W. Australia and West Central	Toodyay (Newcastle)	nanga	mekar	nangar
	Pidong . . .	karong	wilarra	pundarra (*pl.*)
	Minning . .	jinntu	wiyall	burunga (*pl.*)
	Lake Amadeus .	chin-too	peer	pinterry
South of S. Australia and East Central	Narrinyeri .	nunngi	markeri	tuldi
	Parnkalla . .	yurno	pirra	purdli, purli
	Adelaide . .	tindo	piki	purle
	Darling . . .	mengkeeullo	bichooka	boollee
	Diyeri . . .	ditchie	pirra	ditchie thandra
	Murunuda . .	tuna	nanangi	kulaunchi
	Mythergoody .	kumba	goonogoono	ugo
	Larrikeya . .	delirra	lurier	niemellema
North Australia, and Central Australia	Woolna . . .	ummie	loowillea	moorlna
	Daktyerat . .	miru	yuilk	numurul
	Ruby Ck., Kimberley	woloor	yelngong	wurda
	Napier Range, Kimberley	wolgar	bingar	lun
	Sunday Island .	alga	kooîrdi	indi
	Macdonnell Ranges	alinga	taia	quar-allia (W.)
	Walsh R., Q. .	angor	palar	nyolb
	Bloomfield Valley, Q.	woongar	geetcher	mooloowatchur
	Palmer R., Q. .	etha	thargan	ilbanoong
	Coen R., Q. . .	tsche	arroa	ngokoot
	Mapoon R., Q.	ngoa	arroa	ngokwigge
	Gudang, C. York	inga	aikana	onbi
Torres Strait	Kowrarega, Torres St.	gariga	kissuri	titure
	Saibai I., N.G.	gaiga	mölpalö	titoi
New Hebrides.	Aulua Malikula	niel	ambisia	mose
	Nguna . . .	elo	atalangi	ngmasoe
	Aniwa . . .	tera	tumrama	tafatu

Cloud	Sky	Rain	Rainbow
—	—	pokana	weeytena
pona, roona (D.)	—	porrah	wayatih
—	loila (J.)	—	—
—	renn hatara	manghelena	—
bagota (R.)	tooreener (N.)	toorar (N.)	—
lak	wura wura	baan	brinbial
tunmarng	wuruwur	mondar	tyerm
mang	wor wor	wala	derakaworwor
marng	wuruwur	wallah	dherakawurwur
lak	torngor	yayal	—
maing	kotai	midhak	dherake worwor
murnong	—	maiyang	—
munong	munong	maiang	dh'dh'barote
moorn (pl.)	—	kowine	trum
manguay, nurn	trailee	mukaria	worngrie
nort	nguropblindiworak	wflang	wirakalundi
karareyu	—	noorooma	—
—	—	wollong	—
yurong, irawari	wirr, murrumbir	—	gunnung gurran
kurru (plu. R.)	dulka (R.)	panna, wallan (C.)	—
yura, yareil (pl.)	moroko	koiwon	—
gundar, yuro	gunakulla	yuro, kollebari	yulowirri
mundam	nguruindh	yurung	dhakkan
yauggan	bunda	burdoo	—
mara	wolanth	gabby dabut	—
munnditta (pl.)	wallelu	burna	uronguru
mullga-billdi (pl.)	willka	mullga	—
ho-too-worry	ill-carrie	chillberto	—
tuppathauwe	waiirri	parni	kainggi
malko	naieri	kattari, wirra	kuranya
makko	karra ngaiera	kuntoro	kuranye
ninnda, poondoo	korobbyna	mukkra	mondunbara
thularapolkoo	purriewillpa	tulara	kooriekirra
dikura	kunta kunta	nepaulindinka	kudo
ommugoo (?)	murer murer	yappo	kunjo
—	koroa	balmba	—
—	—	mornie	—
berk	anbulk	mada	göndyere
—	piring	nopa	—
	—	nimilar, walar	—
—	—	inra	—
ankata	—	kwatya	umbulara
—	burkuar	kuk	yaman
ngoorpal	tcheari	kapper	yearil
—	—	ogno	—
aveoo	lanna	nuaadhoadhanna	ndrindeni
aveoo	aranra	agaldotanne	andragondhine
otera	—	apura	ung-gebanya
dapar	je	ari	oripara
jia amal	—	ari	—
borinbor	nemar	misa	tiliara
napopouru	nakoroatelangi	usa	asoara
apua	taragi	towa	tumutu

Groups	Dialects	Light	Dark	Morning
Tasmania	East	—	taggremapack	—
	South	—	nune meene larraboo	—
	West and N. W.	—	—	—
	North	—	—	—
	Miscellaneous	tretetea (J.)	—	nigrarua (R.)
Victorian Region	Yarra R.	nguianda	burun	yiram
	Lal Lal	mirriyo	murkal	yirom
	Ercildoune	barp	burun	barp bo barp
	Avoca R.	nyauwi yo	burooin	berip
	Broken R.	yanggim	morporoin	—
	Gunbower	dhyulaipuk	ñarom ñarom	ñuroin
	Warrnambool	nganong	buron	—
	Mortlake	yay, aiap	booboon	neenan
	Booandik, S. A.	—	man kin	—
	Lower Lachlan and Murrum-bidgee	waiange	proandia	tiïa
	Gippsland	mlank	batgalak	wurukamerin
N. S. Wales and S. Queensland	Barwidgee, Upper Murray	torongoro	tiyogolo	—
	Woorajery Tribe. Upper Murray, N.S.W.	—	—	—
	Wiraidhuri	balgangal	burundang	nguronggal
	Turuwul, Port Jackson	—	—	—
	Awabakal	kirrin	—	—
	Kamilroi	turi, burian	nguru (*darkness*)	nguruko
	Kabi, MaryR.Q.	nguruindhau	wuindhau	barbiman
	Warrego R., Q.	durrey	youriugga	buddala
W. Australia and West Central	Toodyay (New-castle)	bena	morrodong	benang
	Pidong	muggerow	mungunnga	mungall
	Minning	—	tunjinnda	—
	Lake Amadeus	chintu-ruigin	moong-a	—
South of S. Australia and East Central	Narrinyeri	nunkulowi	yonguldyi	—
	Parnkalla	birki	ngupinniti	yurno worta
	Adelaide	tindogadla	ngultendi	panyiworta
	Darling	mengkee	wongka	wahmbee
	Diyeri	buralchie	pulkara	manathoonka
	Murunuda	pilpamunina	murra murra	winandinu
	Mythergoody	bertun	kabajee	genool
	Larrikeya	bakuinida	lamingua (C.)	—
North Australia and Central Australia	Woolna	—	lamongwa	—
	Daktyerat	andara	ngö	poïyanggnan
	Ruby Ck., Kimberley	—	—	—
	Napier Range, Kimberley	—	narnbur (*very dark*)	ngimi ngimi
	Sunday Island	—	—	—
	Macdonnell Ranges	alta	albanata	—
	Walsh R., Q.	angor	—	anmun
	Bloomfield Valley, Q.	tyur	ngaul	moannotchi
	Palmer R., Q.	—	ilboong	—
	Coen R., Q.	tscheamboi	doannapini	keammanne
	Mapoon R., Q.	ngoongbai	doannapaini	keammanna
	Gudang, C. York	—	—	—
Torres Strait	Kowrarega, Torres St.	—	—	—
	Saibai I., N. G.	boia	kubil	bataigna
New Hebrides	Aulua Malikula	nutanrien	nuta melingko	veremose
	Nguna	marama	malingo	malikpongi
	Aniwa	merama	pouri	tapopo

DAY	NIGHT	HEAT	COLD
taggre marannye	tagrummena	peooniac	tunack
luggaraniale	nune	lugrab	mallane
loyowibba (J.)	dayna leah	—	—
tridadie	livore	—	—
—	burdunya (P.)	loina (D.)	tenna (D.)
yilinbo	murpuran	wirtnalin	tatirrn
mirriyo	murkal	warwutnyo	munmot
nauwi	burun	katyai	monmut
nyauwi yo	burooin	wiripunya	motilan
karimln	morporondhai	wolorndat	mortarbin
ngolakandok	buroon	wunwundilang	bunbundilang
—	buron	kaluin	kaingeditch
deerung	booroon	kalloon	pallapeek
karo	moal	palawoina (*hot*)	—
nangi	moorprondi	kelali	tinangi
bruindi	bukong	kwarakuan	merbuk
—	—	—	karagutba
—	—	—	polathi
waddu	ngurombang	—	balludai
carmarroo (C.)	purra (R.)	yooroo-ga (C.)	tagora (C.)
purreung	tokoi	—	takara
yeradha	nguru	—	—
nguruindhau	wuindhau	mariman	walai
yoummundurrey	youringgah	cowerly	murnda
bincha	morradong	moonak	netting
muggerow (*daylight*)	mungunnga	kullunngu (*hot*)	murdinnga
womuburrunn	—	—	numulia
—	—	—	war-ringa
nunngi	yonguldyi	waldi	murunkun
marka, yurno	malti	pulla (*warm*)	pai alla
—	ngulti	—	manya
yuko	toongka	boyttyee, kahla kahla	yerkee
kurrurie	pulkara	wuldrulie	kilpalie
orrukuli	murra murra	wiltura	kanini
kumba	kabajing	mundara	yeanga
delirgua	damungua	erringergum	abbulduppi (C.)
irmingn (P.)	—	oorgker	ipegwa
miranbulk	dardarma	poiyadu	mark
—	—	—	—
—	byon	—	—
—	—	—	—
—	igua	—	dana, erinta
angor	anno	—	tan
woongarer	woodjourbu	woombul	bullur
ethuttaga	atha	—	oloorgo
tscheamboi	doannapini	—	taidhömme
ngoakinndi	doannapaini	moipaini	ninnyita
—	yupalga	—	ekanba
—	inur	kanian	sumein
göiga	kubilö	kuamö	sumai
nambung	nuta melingko	kamba kamp	melong kas kas
aleati	kpongi	navivitinuana	namalariana
nopoge	tupo	evera	mukaligi

Groups	Dialects	Fire	Water	Milk
Tasmania	East	tonna	liena	proogwallah
	South	ngune	lia wenee	prooga neannah
	West and N. W.	winnaleah	lia winne	—
	North	padrol	boue lakade	—
	Miscellaneous	une (P.)	mookaria (S.)	—
Victorian Region	Yarra R.	wen	ban	birm birm
	Lal Lal	wīng	ngo pit	pab
	Ercildoune	wī	katyin	kurm
	Avoca R.	wī	katyin	kurm
	Broken R.	wīn	baïn	birm birm
	Gunbower	wonap	kadhin	—
	Warrnambool	wiin	paritch	nga'mo'
	Mortlake	weean	perreech	—
	Booandik, S. A.	—	pare	papainboop
	Lower Lachlan and Murrumbidgee	winapi	kiemie	koimbi
	Gippsland	tauwar	yarn	baag
N. S. Wales and S. Queensland	Barwidgee, Upper Murray	niambunba	—	—
	Woorajery Tribe, Upper Murray, N.S.W.	wongi	kolin	—
	Wiraidhuri	win, guyang	kaling	ngamon
	Turuwul, Port Jackson	gweyong (C.)	bado (C.)	murtin (H.)
	Awabakal	koiyung	kokoin, kulling	—
	Kamilroi	wī	kolle, wollun	—
	Kabi, Mary R Q.	kīra	kung	among
	Warrego R., Q.	wī	nubba	numma
West Central	Toodyay (Newcastle)	kalla	gabby	—
	Pidong	kulla	bubba	—
	Minning	kaiya	kapi, gabi	—
	Lake Amadeus	war-roo	cobbie	—
South of S. Australia and East Central	Narrinyeri	keni	nguk, bareki	ngumperi
	Parnkalla	gadla	kapi kauo	ngamma
	Adelaide	gadla	kauwe	ngammi, ngarru
	Darling	koonyka	nokko	nummaloo
	Diyeri	thooroo	apa	yika
	Murunuda	duro	napa	—
	Mythergoody	yangour	yappo	thambo
North Australia and Central Australia	Larrikeya	kuiangua	karroa	—
	Woolna	letunga	aakie	—
	Daktyerat	tyungo	wawk	gnur
	Ruby Ck., Kimberley	jaba	nopa	—
	Napier Range, Kimberley	—	walar	—
	Sunday Island	nuro	kara	—
	Macdonnell Ranges	ura	kwatya	ilbatya
	Walsh R., Q.	angi	kuk	pip
	Bloomfield Valley, Q.	koongin	banner	bi bi
	Palmer R., Q.	oomar	ogno	oyong
	Coen R., Q.	moi	ngoi	tane
	Mapoon R., Q.	moi	tedi	tane
Torres Strait	Gudang, C. York	yoko	—	—
	Kowrarega, Torres St.	mue	—	ikai
	Saibai I., N. G.	möi	nguki	ikai
New Hebrides	Aulua Malikula	na kamp	nabui	—
	Nguna	na kapu	noai	—
	Aniwa	tiafi	tavai	—

GOD	DEMON	GHOST	SHADOW
—	mienginya	wurrawana	wurrawina tietta
—	ria warrawah noile	riawarawapah	maydena
—	pawtening-eelyle	teeananga winne	belanyleah
—	—	—	—
—	—	—	—
bundyil	ngarang	murup	mula
—	kutyal	murup	mula
bundyil	yulok	waingar	ngark
barnibinmel	natya	murup	ngark
—	ngarang	murup	—
paindyil	ngatta	munganitch	ngarkuk
pirnmaheal	muurup	muurup	wuul'
—	mooroop	boorkoorook	na goe
—	woor, walim	kolandroam	wol
biambule	pongarnoti	popopondi	nouwaki
—	brauwin	mrart	ngauk
—	—	—	—
—	urakabi	—	—
baiamai	baggin	—	gurruman
—	—	mahn (C.)	bowwan (C.)
—	—	mummuya	kommirra
baiame	—	—	—
—	munder	ngutburu	ngutburu
—	—	ninni ninni	gurly gurly
—	—	chinga	—
—	—	—	—
—	—	—	budani
—	mar-moo	—	carn-koo
—	brupe (*devil*)	—	pangari, lilliri
—	marralye	wilya (*spirit*)	madlo (*shade*)
—	kuinyo	towilla	punga, turra
pindee,wahtta,noorinya	boorree	koylppa	kolyppara
moora moora	kootchie	moongara (*spirit*)	—
karina	yarkamata	—	tati
—	—	simbingergolun	mungo
molnganding	berauel	darimiet	—
—	—	—	—
—	bararang	bararang	molang
—	—	—	—
—	nouri	nouri	—
nari	—	—	—
—	erinya	mangabara	undoolya (W.
—	pirkooir	—	baru
—	tchopo	—	wharbur
—	—	inmaningam	—
—	ngai, tschoa	ngai, tschoa	anndormre
—	amvou, tschoa	amvou, tschoa	anndormre
—	—	umboypu	—
—	—	markai	—
augada	markai	—	—
atua	temes	nenanta	nenanta
sukpe	natana sa	natemate	melu
atua	tetua, tiapolo	tetua	noate

GROUPS	DIALECTS	WIND	MIST	SMOKE
Tasmania	East	rawlinna	—	progoona
	South	rallinganunne	—	—
	West and N. W.	lewan	—	—
	North	tegouratina	—	—
	Miscellaneous	linghenar (N.)	—	boorana (R.)
Victorian Region	Yarra R.	munmot	burang	burt
	Lal Lal	winmaling	porang	burt
	Ercildoune	maia	ñura	burt
	Avoca R.	maiya	kairagair	burt
	Broken R.	guron	borang	bort
	Gunbower	miriny	kua	pordok
	Warrnambool	ngonduk	—	thoön
	Mortlake	noondook	wadawoort	dhung
	Booandik, S. A.	niricha	—	booloin
	Lower Lachlan and Murrumbidgee	wilangi	takombi	borti
	Gippsland	kraur	bauindong	dyun
N. S. Wales and S. Queensland	Barwidgee, Upper Murray	—	—	toombaba
	Woorajery Tribe, Upper Murray, N.S.W.	thowera	—	—
	Wiraidhuri	—	gulbi, guang	—
	Turuwul, Port Jackson	gwarra (H.)	—	cudyal (H.)
	Awabakal	wippi	boaring, koropun	poito
	Kamilroi	maier, buriar	dhuber	du
	Kabi, Mary R.Q.	buran	kuang	wului
	Warrego R., Q.	yerga	dunederra	durren
W. Australia and West Central	Toodyay (Newcastle)	nanga	—	kerra
	Pidong	winnju	muluwa	bungu
	Minning	piriddi	—	kaiya-puya
	Lake Amadeus	wolpa	u-bee-terra (fog)	poy-you
South of S. Australia and East Central	Narrinyeri	maiyi	dlomari	kari
	Parnkalla	—	malbara	puyu
	Adelaide	waitpi, warri	madlo	puiyu
	Darling	yertto	poondoo-poondoo	poondoo
	Diyeri	wathara	thoodaroo	ukardie
	Murunuda	chimo	kuinin	kudo
	Mythergoody	murlbunoo	buloothupal	yungoolkar
North Australia and Central Australia	Larrikeya	manmanma	—	kuiugua
	Woolna	minma	—	lemoogiema
	Daktyerat	wurrurk	wen	arabul
	Ruby Ck., Kimberley	—	—	—
	Napier Range, Kimberley	wangool	—	—
	Sunday Island	—	—	—
	Macdonnell Ranges	wurinya	in-jeer-may-jeer (W.)	—
	Walsh R., Q.	—	—	—
	Bloomfield Valley, Q.	moyur	woorpal	gobo
	Palmer R., Q.	olboongol	—	orkoon
	Coen R., Q.	woinji	—	—
	Mapoon R., Q.	tschorita, woinji	—	—
	Gudang, C. York	alba	—	ekora
Torres Strait	Kowrarega, Torres St.	guba	—	tuo
	Saibai I., N. G.	guba	dibagö (fog)	tu
New Hebrides	Aulua Malikula	nelang	veraniet	nakamp basua
	Nguna	nalangi	namavu	nasua
	Aniwa	tumtagi	tefu	tousafi

Thunder	Lightning	Country	Ground
poimettya	poimettye	—	pyengana
papatongune	poimataleena	—	mannina
—	rayeepoinee	—	nattie
—	—	—	longa (J.)
nawawn (D.)	nammorgun (D.)	—	gunta (J.)
ngurndavil	baradyuk	bik	bik
mondara	morinyuk	tyakak	dya
maandar	melarkok	tya	tya
mondara	wilibuk	tyakñak	dya
ngundabil	tyiringingundabil	bek	bek
mondor	dbyulipuk	dhañuk	dha
mondal	—	mering	murang
mundall	yerwun, dherwun	meering	meering
murndal	minanmum	mraad	niraad
mundari	tollpoie	tongi	tongi
kwaran	mlangbitch	wurak	wurak
mundera	narawahnyo	—	—
tumberumba	mikki	bimble	thugoon
murruberai	migge	ngurangbang	dagun
morungle (H.)	manga (H.)	—	pemall (H.)
mulo	malma	—	purrai
tulumi	mi, ngurumi	—	—
mumba	bolla	dha	dha
burdoo	wonning	mye	mye
mulligar	winliding	boojar	boojar
kumurdu	kunde	—	burrna
—	—	—	bana
toney	pin-pan	—	mun-da
munti	nalin, nalurmi	—	tuni
kuranna	—	yerta	yurra
biturro	karndo	pangkarra, yerta	yerta
bootta	kulla-koonyka, berla	geerra	mundee
thularayindrie	tbularakinie	—	mitha
pilpamaninkura	wyinina	pala	pala
roonga	roonga	nargee	nargee
lalluelball (C.)	laurba	—	kuiarloa
leuwee	—	teenger	—
darawiya	gwindyuru	dak	wŏndyö
—	—	—	—
—	—	burra	boorar
tamong	idum	—	kura (*earth*)
—	kwatyabara	mirror (W.)	arila
—	milivir	—	wai
—	—	yamber	bobo
tcharamilli	balpae	—	—
dragette	—	makootte	ogoa
arrokutti	nrepogono	makwigge	ogoa
wagel	omba	—	ampa
duyuma	baguma	laga (*land*)	apa
—	—	—	—
nurvur	nembeli	tipsa	netan
tovae	navila	navanua	natano
tafatibiri	tuptupeia	tageraku	takere

Groups	Dialects	Stone	Hill	Creek
Tasmania	East	loantennina	poimena	—
	South	loinah, lonna	layete paawe	—
	West and N. W.	loine	—	—
	North	lenn parena	meledna	—
	Miscellaneous .	larnar (N.), longa (D.)	neika (D.)	warthanina (J.)
Victorian Region	Yarra R. . .	long	yurndabil	gurnong
	Lal Lal . . .	la	panyal	yalok
	Ercildoune . .	la	kawa	bar
	Avoca R. . .	lakh	kauwa	bor
	Broken R. . .	mordyir	-ngorak	kurnung
	Gunbower . .	laar	ponyul	yalok
	Warrnambool .	morai	—	dhartum
	Mortlake . .	merri	kung, kaa	burang
	Booandik, S. A.	muri	boopik	yaro
	Lower Lachlan and Murrumbidgee	kwiarpi	porporkui	yerani
N. S. Wales and S. Queensland	Gippsland . .	walung	krangark	keauwitch
	Barwidgee, Upper Murray	gibba	bubbura	jeringemor
	Woorajery Tribe, Upper Murray, N.S.W.	gibber	bulga	billabong (*big, dry*)
	Wiraidhuri .	—	bangala	gungan, billa (*river*)
	Turuwul, Port Jackson	kibba (*rock*, H.)	—	turagung (R.)
	Awabakal . .	tunnung	bulka	kirunta
	Kamilroi . .	yarul	taiyul	—
	Kabi, MaryR.Q.	dhake	kunda, tunba	wirra
	Warrego R., Q.	bougal	—	—
W. Australia and West Central	Toodyay (Newcastle)	boyye	katta	billou
	Pidong . . .	murrda	kurrba	wila
	Minning . .	buri	buri	—
	Lake Amadeus .	pulley	—	car-roo
South of S. Australia and East Central	Narrinyeri . .	marti	ngurli	kur (*river*)
	Parnkalla . .	kanya	purri	parri (*river*)
	Adelaide . .	pure	yertamalyo	parri (*river*)
	Darling . . .	gibba	bolo	kulppa, dalyy
	Diyeri . . .	murda	—	—
	Murunuda . .	mudra	wyirira	kuri
	Mythergoody .	mindee	morjo	ooldo
North Australia and Central Australia	Larrikeya . .	belpella	gumarooka (C.)	—
	Woolna . . .	lungea, lunga	lilyerwer	toipunger
	Daktyerat . .	wulu	milgning	yaodyer
	Ruby Ck., Kimberley	pili	—	—
	Napier Range, Kimberley	wirneguni	—	rurar
	Sunday Island	kolb	porit	idal (*river*)
	Macdonnell Ranges	puta	elgata, puta	lara
	Walsh R., Q. .	turn	angguan	algin
	Bloomfield Valley, Q.	kolgi	mungel	yilgi
	Palmer R., Q. .	oolkon	jakkaro	—
	Coen R., Q. .	ogworre	pai	re
	Mapoon R., Q.	ogworre	pai	re
	Gudang, C. York	olpa	pada	—
Torres Strait	Kowrarega, Torres St.	kula	pada	—
	Saibai I., N. G.	kula	pado	kussa (*river*)
New Hebrides	Aulua Malikula	nevit	narah	emburea
	Nguna . . .	vatu	tava	—
	Aniwa . . .	tafatu	aora	teretu, tavai, tafe

GRASS	TREE	BARK	WOOD
rouninna	loatta	poora, poora-nah	wyena
nemone	toronna (D.)	warra	weea
probluah	—	poora leah	—
—	—	—	—
poene (P.)	weena (R.)	moomene (N.)	weenar (N.)
boait	—	darbo	kalk
baran	gur	garong	kalkalk
boaitch	kalk	bam	kalkkalk
bowatch	kalk	mityak	kalk
banom	kalkponyir	yellam	kalk
boatch	pial	muradyap	pial
bodhun	wurutth	muroitch	wiin
karrawan	—	dhurung	weean
boo-tho	—	moorn-dart	wurnap
worlengi	paila	ngorti	pittarkuri
bon	kalak	nondak	kalak
mooro	—	karrayu	toorga
—	yarra (*gum*)	—	keegal
buguin	maddan, gidya	—	maddan, win
bumbur	waddy (H.)	—	waddy (H.)
woiyo	kollai	bukkai	kollai
gorar yindal	tulu	tura	tulu
ban	dhu	kombar	dhu
yowwi	wan	biddal	wi
gilba	bonna	yorda	bunna
bulga	—	—	winnda
—	bulgarra	—	yannda
putta	er-nar	lick-caraka	wor-rue
kaiyi	yape	yorli	lamatyeri
yutara	warndu, wadla	yulti	birka, gudla
—	—	tidli, bakka	gadla
mootto	koombahla (*gum*)	tulkeroo, pultta	yerra
kuntha	—	pitchie	thooroo, pitta
kuntabukra	wewa	wita	turo
kutthree	bargour	simbe	bargour
merelma	mardpurma	mangguruma (C.)	marriburma (C.)
lugilyer	—	leemoconana	meurwer
weno	tyungo	duil	wundallo
yuka	—	—	—
wooroo	—	—	bonar
koorlyo	burduch	—	—
nama	—	bailka, irknala	rula
alku	iringkin	harun	angi (*log*)
karer	choko	toombul	choko
ookin	—	oonkil	oomar
lainne	—	kaii	ko
lainne	—	kago	ko
untinya	pure	ranga	yoko
burda	prue	purur	mue
hupö	kaipui	pia	pui
namine	naki	nakalukte naki	naki
nangmenau	nakau	nawili nakau	nakau
tagaferi	terarakou	nokiri terarakou	terarakou

Groups	Dialects	Camp	House	Hole
Tasmania	East	—	lenna	—
	South	—	line	—
	West and N.W.	—	—	—
	North	—	—	—
	Miscellaneous	—	leebrerne (N.)	—
Victorian Region	Yarra R.	willam	willam	bakporn
	Lal Lal	karong	karong	mir
	Ercildoune	laa	maiam maiam	baam
	Avoca R.	lar	lar	mir
	Broken R.	yellam	yellam	miring
	Gunbower	lar	lar	ludhuk
	Warrnambool	wurn	porpornduk	pakweitch
	Mortlake	woor	burburnonook	ganung
	Booandik, S. A.	noorla	ngoorla	—
	Lower Lachlan and Murrumbidgee	lingingi	kundi	ngurngi
	Gippsland	bong	katyun	ngang
N. S. Wales and S. Queensland	Barwidgee, Upper Murray	gunya (?)	—	—
	Woorajery Tribe, Upper Murray, N.S.W.	—	gunya (?)	—
	Wiraidhuri	—	gawier	milbi, munil
	Turuwul, Port Jackson	gonyi (*hut*, H.)	kunje (*hut*, R.)	gomira (H.)
	Awabakal	koiyong	kokere	—
	Kamilroi	—	kundi	—
	Kabi, MaryR.,Q.	kīra, kīrami	tura	nolla
	Warrego R., Q.	nurra	gundi	wordu
W. Australia and West Central	Toodyay (Newcastle)	kalla nanup	mia	—
	Pidong	—	ngora	igil
	Minning	—	mullga-minntoradd	—
	Lake Amadeus	—	—	—
South of S. Australia and East Central	Narrinyeri	manti, ngauandi	karuturi	merki
	Parnkalla	yurla, karnko	karnko	tyeka yappa
	Adelaide	—	wodli	tau, yappa
	Darling	yuppara	goollee	meengga, woollee
	Diyeri	oora	boonga	willpa
	Murunuda	ura	tua	mikri
	Mythergoody	magier	yinbar	kornjar
	Larrikeya	guinidirk	manolürra	gauga
North Australia and Central Australia	Woolna	wylie, mikehr	—	wawee
	Daktyeral	dak	anduk	yalo
	Ruby Ck., Kimberley	—	—	—
	Napier Range, Kimberley	yalbah	miar	nimilar
	Sunday Island	ooroorunggari	—	—
	Macdonnell Ranges	tmara	ilta, lunga	altyura
	Walsh R., Q.	alpa	polkan	irpen
	Bloomfield Valley, Q.	bulpa	bian	tchanko
	Palmer R., Q.	ogue	—	—
	Coen R., Q.	nge	wutschu	—
	Mapoon R., Q.	nge	wutschu	—
	Gudang, C. York	—	eikuwa	—
Torres Strait	Kowrarega, Torres St.	mudu	laga	—
	Saibai I., N.G.	—	lagö	arkatö
New Hebrides	Aulua Malikula	vere (*village*)	nimwa	nambul
	Nguna	—	nasungma	moru
	Aniwa	—	tafare	terua

LUMP	PATH	FOOTMARK	THE BLACKS
—	—	puggataghana	—
—	—	pallowa lugganah	—
—	—	pah lug	—
—	—	—	—
—	—	—	—
jiruin	baring	—	kulin
turung	kandor	kandor	koli
karinga	baring	baring	koli
turung	baring	barkuk	kuli
yulabil	baring	baringidyinang	kulin ba bedyir
dhunkauir	parin	parin	kuli
koreda	taan	—	maar
mamit	dhaarn	dhinnaneong	maara
—	ware	teena	—
tupatupaule	laimbi	thinangi	woongi
dhanbilan	wanik	wanik	konai
—	karrika	—	—
—	—	—	—
buabuawanna	yabbang	bai, darrambal	main
—	—	—	eora
bulka	yapung	yulo	—
—	turabul	—	murri (*s.*)
wulbo	kuan, ulu	kuan	dhan
curlewru	yourroun	dinnadonybu	myeing (*maiīng*)
—	—	genna	yunger boola
—	—	—	—
—	—	kulbia	—
—	u-worra (*road*)	—	—
—	yarluke	yarluki	uarrinyeri
bakkarra	widla	—	—
—	tappa	tainga	—
—	tinna	tinna	wimbaja
—	—	thidna	kurnawara
tunka	waruwaru	tinbuto	kurna
nambul	wathoo	janner	—
—	kuiatburroa	beaitbar (C.)	belirra
—	—	ya-wehrl	—
povo	eru, widbeldyerang	mel	gnan
—	—	—	—
—	gurdy, karty	karty, nimblar	—
—	—	—	—
wola (*heap*)	tyaia, tnalta	—	—
—	tel	tel	pama
tchungi	boral	boural	bummer
—	—	amul	—
patoo	tave	kwe	mrittakke nambarra
poi	rago	kwe	mritakke nambarra
—	—	—	—
—	—	—	—
—	yabugudö	—	—
—	napun	nangmele	natangmoli loa maga
—	havila	meleluan	asamangk miet
lampe (?)	teretu	tumavai	tagata pouri

Groups	Dialects	A Blackfellow	A Blackwoman	Man
Tasmania	East	pugganna	lowanna	pugganna
	South	pullawah	lowanna	pallawah
	West and N.W.	pah-leah	nowaleah	pah-leah
	North	—		looudouene
	Miscellaneous	—	louana (R.)	wibar(N.)penna(D.)
Victorian Region	Yarra R.	kulin	baigurk	kulin
	Lal Lal	koli	bagurk	koli
	Ercildoune	koli	bangbanggo	koli
	Avoca R.	kuli	baibago	kuli
	Broken R.	kulin	bedyir	kulin
	Gunbower	bang	leiruk	wotok
	Warrnambool	maar	dhanambul	maar
	Mortlake	mara	dhumdhumboorn	mara
	Booandik, S. A.	drual	kinekine-nool	drual
	Lower Lachlan and Murrumbidgee	woongi	liorki	kolkorni
	Gippsland	konai	wurukot	konaiwadi
	Barwidgee, Upper Murray	—	—	jirri
	Woorajery Tribe, Upper Murray, N.S.W.	mian	—	—
	Wiraidhuri	main	—	gibbir
N. S. Wales and S. Queensland	Turuwul, Port Jackson	mul-la	din	—
	Awabakal	—	—	kore
	Kamilroi	murri	yīnar	giwīr
	Kabi, MaryR.Q.	dhan	yīran	kīvar
	Warrego R., Q.	myeing	mugging	mying
W. Australia and West Central	Toodyay (Newcastle)	yunger moon	yungar yago moon	mamerup
	Pidong	—	—	yammeji
	Minning	—	—	minninng
	Lake Amadeus	bar-too (*man*)	ho-carra	bartoo
South of S. Australia and East Central	Narrinyeri	korni	mimini	korne
	Parnkalla	—	pallara	miyerta, yura
	Adelaide	—	—	meyu
	Darling	wimbaja	burrukka	wimbaja
	Diyeri	kurna	widla	—
	Murunuda	kurna	punga	karuro
	Mythergoody	moorey	pungar	—
	Larrikeya	barning (C.)	barning-ceimcur (C.)	molinyo
	Woolna	looarkieinga	mungedma	kumaol (*mankind*)
	Daktyerat	gnan	elugur	gnolan
North Australia and Central Australia	Ruby Ck., Kimberley	—	—	wunwa, papa
	Napier Range, Kimberley	wambar	—	—
	Sunday Island	—	—	amba
	Macdonnell Ranges	—	—	atua, erila, atula
	Walsh R., Q.	moak	yerkul	lunjin
	Bloomfield Valley, Q.	bummer	tchalbu	dingar
	Palmer R., Q.	immi	aruntha	—
	Coen R., Q.	nambarra	andrommre nambarra	adhetroo
	Mapoon R., Q.	nambarra	andrommre nambarra	annege
	Gudang, C. York	amma	—	unbamo
Torres Strait	Kowrarega, Torres St.	garkai	—	turkekai
	Saibai I., N. G.	—	—	mabaigö
New Hebrides	Aulua Malikula	natangmoli loa sikai	nangoroi loa sikai	natangmoli
	Nguna	asamangk miet	tambaluk miet	teta, asamangk
	Aniwa	tase tagata pouri	fafine pouri	tatane, tane

OLD MAN	WOMAN	OLD WOMAN	BOY
lowlobengang (J.)	lowanna	payana	cotty-mellitye
—	lowanna	nena ta poinena	poilahmaneenah
—	nowaleah	—	—
—	loubra	—	—
lalubeguna (L.)	lurga (D.) lolna (J.)	lowla pewanna (J.)	leewoon (D.)
wikabil	baigurk	murndigurk	bobup
didabil	bagurk	mundagurk	golkonkuli
mati koli	bangbanggo	mati bangbanggo	golkongolkon
ngambin	baibago	ngunyimgoork	bupup
dbaingula	bedyir	wirk wirk	bobopdhak
wanyim	rembindyuk, leyruk	wunyimkurk	bupang
ngolla	kuguwitch	ngarom ngarom	waran waran
alungalung	dhumdhumboorn	koowoowetch	warun warun
ngiring gee	kinekine-nool	porpegngara	koonatgo
pokongi	moroingham	kuambiliki	piangongi
budhan	wurukot	kwerailmina, wurukot	lidh
jirribong	jeri	kauwantigba	uaro
jeeribung	—	jeeribung (?)	boori
dirribang	inar	ballagun	biran, burai
tangung (R.)	—	mulda (R.)	wongerra (C.)
ngarombai	nukung	ngarongeen	—
diria (*old*)	yinar	yambuli	birri
winyir	yiran	marun	nguin
mutchaburry m.	mug-erding	burraka	yungurd
mongan	yago	billing	kooling
winnja	wanndi, nyalo	thukurrn	murdin
—	wurnanng	—	yina
—	ho-karra	koon-ja-gilbee	oll-ar
yandiorn	mimini	yandi-imin	—
bulka, kulya	pallara, ngammaityu	—	mambarna
—	ngammaitya	paityabulti	kurkurra
mertta	burrukka	nahnggo	willya-roonga
pinaroo	widla	wildapina	kurawulie
karuro mutchuebu	kuei	kuei mutchucha	wei
more	—	womoora	jueary (?)
lariba	onullaga	goomool (C.)	nim, nemerk
longailinga	mungedma	iteburna	—
bork	wundinigmun	mirmallo	notyur
—	aringa, nama	—	—
darral	ihandur	—	—
—	oorang	—	—
ayna	—	—	wora
piringga	wolnga (*young*)	tombi	wurkun
pinga	tchalbu	kumber	warru
oolpa	—	—	—
wattepoi	dronnanne	andorpatroo	pfoimakonne
waggapoi	andrommre	andorprigge	fopparri
—	undamo	—	—
keturkekai	ipikai	kefpikai	turkekai kaje
moroigö	ungwakazi	—	magina kazi
natangmoll matua	nangoroi	nangoroi matua	piakiki nanoai
teta nurseramp	tambaluk	tambaluk nurseramp	tamare
tatane ituai	fafine	fafine itua	tatane sisi

Groups	Dialects	Girl	Baby	Father
Tasmania	*East*	lowana keetanna	cottruluttye	noonalmeena
	South	longatyle	puggata riela	nanghamee
	West and N. W.	—	rikente	tatana (J.)
	North	—	looud	nimermina
	Miscellaneous .	ludineny (J.)	badany (J.)	munlamana (J.)
Victorian Region	*Yarra R.* . .	murnmurndik	bobup, waiyibo	maman
	Lal Lal . . .	madamundik	bupup	bitang
	Ercildoune . .	bunai bunai	bupup	maamuk
	Avoca R. . .	bunya bunya	nilamgurk	mamuk
	Broken R. . .	bornai	bubupdhark	mamano
	Gunbower . .	wadhibuk leyruk	bupang	mamuk
	Warrnambool .	paraiparaitch	bupop	bipai
	Mortlake . .	buriburetch	boopoop	pepie, beebie
	Booandik, S, A.	barite, koonam	—	marm
	Lower Lachlan and Murrumbidgee	maiwangupi	popopi	mamoma
	Gippsland . .	dhalu wurukot	dhaliban	monggan
N. S. Wales and S. Queensland	*Barwidgee, Upper Murray*	weki	—	—
	Woorajery Tribe, Upper Murray, N.S.W.	beelarjerod	—	mooma
	Wiraidhuri .	inargung	munga	babbin
	Turuwul, Port Jackson	werowey (C.)	nabungay (C.) wuidalliez	beanga (H.)
	Awabakal . .	murrakin	bobog	biyungbai, bintun
	Kamilroi . .	mie	kaingal	buba
	Kabi,MaryR.,Q.	wuru	wolbai	pabun
	Warrego R., Q.	gunney	gudderra	budding
IV. Australia and West Central	*Toodyay (Newcastle)*	yago kooling	nobain	maman
	Pidong . . .	thura	maiu	mumma
	Minning . .	—	—	—
	Lake Amadeus .	cue-on-buntor	—	cartoo
South of S. Australia and East Central	*Narrinyeri* . .	bami	kelgalli	nanghai
	Parnkalla . .	kardni	kaitya	pappi
	Adelaide . . .	mankarra	tukkutya	yerlimeyu
	Darling . . .	pulkahly	kichungga	kahmbeeja
	Diyeri . . .	koopa	koopa	apirrie
	Murunuda . .	kuei	pula	napira
	Mythergoody .	mungane	churloo	yadthoo
	Larrikeya . .	maneiga	larree (C.)	neganbira
North Australia and Central Australia	*Woolna* . . .	—	—	bipie
	Daktyerat . .	windyarello	mölmülma	ngaolu
	Ruby Ck., Kimberley	—	—	ombuna
	Napier Range, Kimberley	yabun	baba	yabellar
	Sunday Island .	—	—	—
	Macdonnell Ranges	kwara	um-bra-coora (W.)	kata, nekua
	Walsh R., Q. .	yerkul	tapu	undya
	Bloomfield Valley, Q.	maral	ouar	andgan
	Palmer R., Q. .	—	awillung	atheem
	Coen R., Q. .	morgatanne	agame	naita
	Mapoon R., Q. .	lande, marprenne	agame	naita
	Gudang, C. York	—	—	epada
Torres Strait	*Kowrarega Torres St.*	ipikai kaje	muggi kaje	baba, tati
	Saibai I., N. G.	ngawakazi	—	babö, tati
New Hebrides	*Aulua Malikula*	piakiki nangoroi	pipia	tamana
	Nguna . . .	tambaluk kakas	anetana (*child*)	teta
	Aniwa . . .	fafine sisi	tama sisi	tata

Mother	Husband	Wife	Elder Brother
neingmenna	puggan neena	—	—
neemina	pah-neena	—	—
neena moygh	—	—	—
blemana	—	—	—
powamena (J.)	—	cuani (P.)	—
bapa	nangguronga	brimbanna	banggon
ngatonyuk	nanabun	nganapunyuk	wardunyuk
bapuk	anyetyuk	tyaptyapuk	waruk
papuk	anityuk	nataguruk	wawuk
babono	nanggorong	bimbarno	banggono
pabuk	anidyuk	motminyuk	wawuk
ngira	ngonabun	malongar	wardai
yurungi	anaboort	mulladh	wartietsh
ngat	nganap	mala	wargul-e
korna	nopa	lileli	mouri woni
yakon	benong	laua	dhandon
—	—	bularjeru	—
hubba	—	—	kokong (*brother*)
—	mammadin	mammadin	gagang
wyanga (H.)	—	maugobn (C.)	babunna (C.)
tunkan	poribai	porikunbai	bingngai
ngumba	gulīr	gulīr	daiadi
ngavang, ubung	dhandor	malimgan	nuin
kiah	girring	nubba	bauīng
jukan	mamon	kardo	kardijet
—	murrdong	murrdong	kurda
yack-hoo	coo-rie	coo-rie	coota (*brother*)
nainkowa	nape, napalle	nape	gelanowi
ngammi	yerdli	karteti, yungara	yunga
ngankitta	ngubba	karto, yangarra	yunga (*brother*)
nunmaka	mahlee	koombahka	kahkooja
andril	noa	noa	niehie
narmidi	—	—	—
meerala	narthia	narthia	arboon (*brother*)
kuding	—	aladig	qualaliva (C.)
kardie	lainelongunya (P.)	—	nulla
gnagnaain	nengl	elngen	pukang
nume	—	wompan	—
kooya	—	—	ba-bellar
—	malardin	irwinya	koira (*brother*)
maia, makura	noa	noa	uckfllya (W.)
hauwa	moa	wau	uping
ngarmo	tchoniu	munyur	yapper
among	—	—	athil
tatoo	ngonoongbrange	ngonoome	manaen
adhai	ngioongbrange	ngiamre	maianne
atinya	anba	onda	—
anima, apu	allai	ipi	adoama (*brother*)
ama, apu	imi	imi	kuikuiga
kpilana	anawota	anangoroi	taina takalapa
nina	asunu	asunu	yeye
mama	nunwane	inahune	noso sore

Groups	Dialects	Younger Brother	Elder Sister	Younger Sister
Tasmania	East	—	—	—
	South	—	—	—
	West and N.W.	—	—	—
	North	—	—	—
	Miscellaneous	—	—	—
Victorian Region	Yarra R.	dhidhidha	langtang	dhidhidha
	Lal Lal	wangatuk	tatyurun	bormborañuk
	Ercildoune	kortuk	tyatyuk	kortyuk
	Avoca R.	kutuk	tatyuk	kutruguk
	Broken R.	barnan biyu	lathgon	banbonok
	Gunbower	gutmenyuk	dhatyuk	wutenyuk
	Warrnambool	kogu	kakai	kokulir
	Mortlake	marti	kaki	kookeer
	Booandik, S. A.	nere, doate	date	nere-er
	Lower Lachlan and Murrumbidgee	maimi woni	mouri tati	mainni ki
	Gippsland	dhalitch	boandhan	landok
N. S. Wales and S. Queensland	Barwidgee, Upper Murray	—	—	—
	Woorajery Tribe, Upper Murray, N.S.W.	—	nıyngan (*sister*)	—
	Wiraidhuri	gulmain	mingan (*eldest*)	muogan
	Turuwul, Port Jackson	—	mamunna (G.)	—
	Awabakal	kumbul	—	—
	Kamilroi	gulami	boadi	bure
	Kabi, Mary R., Q.	wuthung	yabun	naibar
	Warrego R., Q.	mourn	bubba	genyera
South of S. Australia and East Central	Toodyay (Newcastle)	kootadong	katamity	katamity
	Pidong	bua	judo (*sister*)	bua
	Minning	—	—	—
	Lake Amadeus	—	kongaroo (*sister*)	—
	Narrinyeri	tarte	maranowi	tarti
	Parnkalla	ngaityaba	yakka (*sister*)	—
	Adelaide	—	yakkana (*sister*)	—
	Darling	bahlooja	wahttooka	wahtteeja
	Diyeri	athata	kakoo	athata
	Murunuda	—	—	—
	Mythergoody	—	moona (*sister*)	—
North Australia and Central Australia	Larrikeya	mineemiller (C.)	buerra (C.)	jeramooka (C.)
	Woolna	wetter	nelami (P.)	wetter, wutta
	Daktyerat	nundang	aldang	nundang
	Ruby Ck., Kimberley	—	—	—
	Napier Range, Kimberley	mallar-kartin	—	—
	Sunday Island	—	gwira (*sister*)	—
	Macdonnell Ranges	itia	urumba	itia
	Walsh R., Q.	uping	abar	hongark
	Bloomfield Valley, Q.	yapputchiu	barbar	ginkiur
	Palmer R., Q.	amoko	thuppa	ejeeja
	Coen R., Q.	otroo	kwitte	otroo
	Mapoon R., Q.	tanoombanne	loege	tanoombanne
	Gudang, C. York	—	—	—
Torres Strait	Kowrarega, Torres St.	—	—	—
	Saibai I., N. G.	kutaiga	babatö (*sister*)	—
New Hebrides	Aulua Malikula	taina takariki	ngorena takalapa	ngorena takariki
	Nguna	atisina	rabina	—
	Aniwa	noso sisi	nokave sore	nokave sisi

CHILDREN	CANOE	FISH	PORCUPINE
—	mallanna	mungunna	mungyenna
bewoon (*sing.* J.)	nunganah	peeggana	mungye
—	nunghuna	—	mungynna kanagale
looweinna (J.)	—	pinounn	milma (J.)
pugyta (*sing.* R.)	lukrapani (L.)	penunina (J.)	trewnina (J.)
wurtona	kurong	towet (*blackfish*)	kauwarn
burunbalok	yogoip	worapin	monngark
karkar bupup	bam	yurtyuk	yulawil
bupupkalik	yolgoip	wirap	yulawil
bonbonarik	korong	malon	kauworn
bupang	yungutch	wiringal	lipkil
tukui tukui	dhurung	—	wilangalak
dhoie dhoie	dhorong	erigar	goonama
	—	—	
pangongi	yongopi	parndeli	yerendingi
yenilmin	girl	kain	kauon
—	doothoo	yumbo	deino
—	murring	munji	
burai (*s.*)	—	guya	—
goroong (C.)	nowey (C.)	magra (H.)	—
wonnai	nauwui	makora	—
kai	kumbilgal	guiya	—
kumma, wolbai	kombar	bala, undaiya	kakkar
gudderrakulgra	kunarew	kewya	—
koolong	—	kalbart	—
—	—	wabi	kundiwa
—	—	—	—
—	—	un-dippin	—
porlar	meralti	mami	—
—	karnkurtu	kuya	—
wakwakko	—	kuya	—
berloo-berloo	booltaroo	tahpooroo, perndoo	kultoo
koopawura	pirra	paroo	—
—	—	—	
churloo	nangool	palby	wychguine
nemebira	gunoogara (C.)	maddo	—
—	moërty	liyer, lieya	—
bulk	wendu	dugur	menak
yambadi (*sing.*)	—	—	—
—	gundig bourough	pee	—
—	—	ali	—
—	—	irbunga (*pl.*)	in-nar-ling-er (W.)
nyolba	—	yu	—
yierke-yierker	marakan	kuyu	ngunkin
—	—	oyi	—
—	patr	nia	—
mrittakke agame	patr	nia	—
—	angganya	wawpi	—
—	gul	wawpi	—
maginakazi (*pl.*)	gulii	wapi	—
piakirikiri	naki	nika	—
tamare	rarua	naika	—
erecriki	tavaka tagata	cika teika	—

Groups	Dialects	Native Bear	Native Dog	Kangaroo
Tasmania	East	—	—	lyenna
	South	—	—	lena, tarrana
	West and N.W.	—	—	ku leab, tarr leah
	North	—	moukra (*dog*)	taramei
	Miscellaneous	—	—	terrar (N.)
Victorian Region	Yarra R.	korbura	wiringwilam	koim
	Lal Lal	ngarmbulamoom	tarwal	koim
	Ercildoune	badyomom	wilkar	kora
	Avoca R.	botyunmum	wilkar	kura
	Broken R.	korbil	wiringalam	maram
	Gunbower	ngarmbulmum	wirangan	kura
	Warrnambool	winggal	pornang	kurai
	Mortlake	wirringill	burnung gaal	goroite
	Booandik, S. A.	—	karnachum	koraa, mare (*f.*)
	Lower Lachlan and Murrumbidgee	—	terilumbi	kuarangi
	Gippsland	kula	ngurain	dhīra
N. S. Wales and S. Queensland	Barwidgee, Upper Murray	—	wehnga	boodjoo
	Woorajery Tribe, Upper Murray, N.S.W.	—	merri	womboin
	Wiraidhuri	—	karingale	bandar
	Turuwul, Port Jackson	—	waregal (H.)	patagurang (H.)
	Awabakal	—	worrikul, yuki, mirri (*f.*)	moane
	Kamilroi	—	murren, maiai	bundar
	Kabi, Mary R., Q.	kulla	wiyidha karum	marī
	Warrego R., Q.	—	yewgee	kulla
W. Australia and West Central	Toodyay (Newcastle)	—	dordoyakain	yunkar
	Pidong	—	duthu (*dog*) waiul	malu
	Minning	—	doychu	pirkuda
	Lake Amadeus	—	—	mar-loo
South of S. Australia and East Central	Narrinyeri	—	turütparni, kel	wangami, tulatyi
	Parnkalla	—	kurdninni	kadlukko
	Adelaide	—	warrukadli	kunda, nanto
	Darling	—	poolkeja	tultta
	Diyeri	—	kintalo (*dog*)	chookaroo
	Murunuda	—	tala	kura
	Mythergoody	—	yamby	maijumba
North Australia and Central Australia	Larrikeya	—	meelinga (C.)	melulla
	Woolna	—	illaya	marning-an-anya
	Daktyerat	—	barundiru	modth
	Ruby Ck., Kimberley	—	—	jugi (or) dudi
	Napier Range, Kimberley	—	yallar	garabil
	Sunday Island	—	ela	piru
	Macdonnell Ranges	—	i-rinka (W.)	rera
	Walsh R., Q.	mungar	tok (?)	amui
	Bloomfield Valley, Q.	—	ngarnio muramo	wallur
	Palmer R., Q.	—	oota	innar
	Coen R., Q.	—	orke	'mvokoo
	Mapoon R., Q.	—	oa	angai
	Gudang, C. York	—	ing-godinya (*dog*)	epama
Torres Strait	Kowrarega, Torres St.	—	umai	usur
	Saibai I., N. G.	—	umai (*dog*)	usaru
New Hebrides	Aulua Malikula	—	—	—
	Nguna	—	—	—
	Aniwa	—	kuri, kuli	—

Opossum	Emu	Iguana	Eaglehawk
lowowyenna	punnamoonta	—	gooalanghta
leena	ngunannah	—	weelaty
papnoolearah	—	—	—
—	—	—	cockinna (J.)
wolimmerner (N.)	pandanwoonta) S.)	leenar (N.)	neirana (R.)
walert	poralmil	bujing	bunjil
walart	karwir	tyulin	ngaromgar
wila	yauwir	wirmbil	werpil
wile	yauir	tyulin	narail
walert	—	tulin	bundyel
wila	kauwir	dhulin	wirbil
kuramuk	kaping	yuruk	ngianggar
kooramook	barinmall	urook	neeungura
kooramo	kower	—	ngeere
pondandi	thungati	wainbali	waiapili
wadhan	maiyor	badhalok	kuanamurung
burra	murriawa	goorooda	wannomurra
willi	pettabang	—	mulyan
willei	ngurain, dinnawan	girua	ibbai, mullian
kuruera (R.)	marry-ang (H.)	—	—
willai	kongkorong	—	wirripang
mute	dino-un	duli	mullion
ngarambi	nguruindb	warui	buthar
googie	nurrung	burnna	kurra
koomal	wagie	mulliwa	walga
—	yallebiri	wadebi	warrida
—	tula	galka	—
wye-hoota	cur-lier	ween-dug-a	wol-lowra (*eagle*)
piltari	pinyali	tiyauwe	wulde
pilla	warraitya	—	yarnu
—	kari	pundonya	pilla, wilto
yarinjy	kulttee	tarkooloo	bilyahra
pildra	woroocathie	kopirrie	curawura
balu	warukatchi	kurininga	witchuhankura
kargoin	jungoobury	muinbooaberry	goorithilla
macmilla (C.)	langura (C.)	kurara (P.)	—
—	moraqunda	—	—
wiyi	ngurin	tyang	murmello
—	—	—	—
lungar	kuriningara	—	—
—	—	—	—
antina	ilia	ad-june-pa (W.)	eritya
adel	pur	konjil	—
yowere	—	tcbatti	yelngur
oolon	oorooba	—	—
—	—	'nrerandre	toarri
—	—	'nreranji	toarri
—	nichulka	—	—
barit	—	—	agaleg
—	—	tamoi	—
—	—	—	—
—	—	—	mala
—	—	—	maruke, nifatu

Groups	Dialects	Crow	Black Duck	Wood Duck
Tasmania	East	lietenna	—	—
	South	taw wereiny	—	—
	West and N.W.	—	—	—
	North	trenn houtne	—	—
	Miscellaneous	kella (L.)	—	—
Victorian Region	Yarra R.	wang	dulum	—
	Lal Lal	wa	tulum	worowirt
	Ercildoune	wa	ngari	biabiarp
	Avoca R.	wa	ngari	beyapyirp
	Broken R.	wang	dolong	baitmom
	Gunbower	waa	ngori	ngonok
	Warrnambool	waa	dhurubarang	—
	Mortlake	wah	thorbrun	naook
	Booandik, S. A.	wa	purner	—
	Lower Lachlan and Murrumbidgee	whalakeli	tolomi	naari
N. S. Wales and S. Queensland	Gippsland	ngarukol	wrang	naak
	Barwidgee, Upper Murray	wagara	dooloomoo	ngaru
	Woorajery Tribe, Upper Murray, N.S.W.	waggan	poothenbong	ngaru
	Wiraidhuri	waggan	wambuain tang	—
	Turuwul, Port Jackson	wagan (H.)	yoorongi (C.)	—
	Awabakal	wakun	—	—
	Kamilroi	waru, waun	karangi	ngurapala
	Kabi, MaryR., Q.	wowa	nar	nar
	Warrego R., Q.	woggan	mirri-nurra	mirri-nurra
W. Australia and West Central	Toodyay (Newcastle)	wardang	munnie	wonda
	Pidong	kaku	—	—
	Minning	—	ngarrawa	—
	Lake Amadeus	carn-ka	chip-pia (*duck*)	—
South of S. Australia and East Central	Narrinyeri	marangani	nakkari	wanye (*mountain*)
	Parnkalla	wornkarra	—	—
	Adelaide	kua	—	—
	Darling	wahkoo	mingara	koonahly
	Diyeri	kowulka	chippala	koodnapina
	Murunuda	wakuri	muto	milkipulo
	Mythergoody	womarine	koopery	alowan
North Australia and Central Australia	Larrikeya	quagabar (C.)	benaymara (C.)	—
	Woolna	—	lermawal	—
	Daktyerat	wangur	pulnirik	—
	Ruby Ck., Kimberley	—	—	—
	Napier Range, Kimberley	wingard	—	chibile (*whistling*)
	Sunday Island	—	—	—
	Macdonnell Ranges	ngapa	—	—
	Walsh R., Q.	ada	kuir	—
	Bloomfield Valley, Q.	—	—	borok
	Palmer R., Q.	atha	onoogi	—
	Coen R., Q.	augaritti	moikapoome	—
	Mapoon R., Q.	rarri	moiboome nambarra	—
	Gudang, C. York	—	—	—
Torres Strait	Kowrarega, Torres St.	—	—	—
	Saibai I., N. G.	—	baga (*duck*)	—
New Hebrides	Aulua Malikula	—	—	—
	Nguna	—	—	—
	Aniwa	niau ouri	togarei ouri	tagarei agarakou

Wild Turkey	Pelican	Laughing Jackass	Native Companion
—	treeoonta langta	—	—
—	toyne	—	—
—	—	—	—
—	treoute	—	—
—	lanaba (J.)	—	—
—	wajil	kurung kurung	kurork
toraiwil	bortangil	kuark	borangeit
kolabityin	patyangal	kowark	gutyun
kulabityin	bortyangil	kuark	norakuang
—	—	gurng gurng	kurork
ngorau	linanggur	koorng koorng	koodhoon
—	—	kunatth	kurun
barim-barim	gardbarup	koonitthe	kooroon
laa	parangal	kooartung	wandi
nuwe	ninangure	kowari	toorkuangi
—	buran	kuak	kurakan
nungarawa	goolakgahle	—	brolga
komether	—	kookaburra	berralgen
gambal	—	—	buralgang
—	car-rang-a-bomurray(C.)	gogannegine (C.)	—
—	karong-karong	—	—
burowa	guleale	kukuburra	buralgha
kalarka	bulla wallum	kawung	kunduran
geeyerra	—	kuggangurru	kurruru
bibilar	—	—	—
—	—	—	—
—	—	—	—
—	—	—	—
talkinyeri	nori	—	prolggi
walla	widli	—	—
wolta	yetu	—	—
tikkara	booleeja	korrookkahkahka	goolerkoo
kulathoora	thaumpara	—	boo·alkoo
wankinara	malimaro	—	—
thurua	walkuperry	jarungodl	toorka
lamamu (C.)	madaridja (C.)	lanurba (C.)	toluba (C.)
—	lourpita	kargak	—
yuntyumunur	monongur	kulbobuk	elinyunung
—	—	—	—
—	—	—	—
—	—	—	—
itoa	kabilyalkuna	—	—
—	—	—	—
waloroongur	piluara	wabougoka	kourpal
—	atharoo	—	ingibbi
yambanyi	adhaurotte	anjomme	pronjomme
yambinne	adhaurootte	anjomme	prondorme
araunya	—	unbunya	aporega
raon	—	kowon	aporega
surka	—	—	—
—	—	—	—
malau	—	—	—
manu sori	—	—	—

Groups	Dialects	White Cockatoo	Black Cockatoo	Swan
Tasmania	East	weeanoobryna	menuggana	kelangunya
	South	nghara	nghay rumna	pugheritta
	West and N.W.	—	—	publee (J.)
	North	—	—	cocha (J.)
	Miscellaneous	—	—	katagunya (J.)
Victorian Region	Yarra R.	ngaiyuk	ngyernong	kunuwara
	Lal Lal	dyinap	wiran	kunuwar
	Ercildoune	dyinyap	weran	konawar
	Avoca R.	dyinyap	wiran	kunuwar
	Broken R.	ga'an	yenggai	kunuwara
	Gunbower	tyinap	wiral	kunuar
	Warrnambool	ngayuk	wilan	kunuwar
	Mortlake	nayuk	woolan	koonawarra
	Booandik, S. A.	mar, karaal	wiler, treen	koonowor
	Lower Lachlan and Murrumbidgee	karandi	—	tanapuki
	Gippsland	braak	ngirnak	kitai
N. S. Wales and S. Queensland	Barwidgee, Upper Murray	gadauna	neanyo	mullewa
	Woorajery Tribe, Upper Murray, N.S.W.	waima	—	goonak, thunthu
	Wiraidhuri	murrain	billir	—
	Turuwul, Port Jackson	garaway (H.)	carall (H.	mulgo (C.)
	Awabakal	kearapai	wailla	—
	Kamilroi	biloela	—	burunda
	Kabi, Mary R., Q.	gigum	wiyal	kulun, ngiring
	Warrego R., Q.	—	—	—
W. Australia and West Central	Toodyay (Newcastle)	munait	—	—
	Pidong	bulli	biddiarra	kurilithu
	Minning	—	—	—
	Lake Amadeus	cock-a-lella	yeranda	—
South of S. Australia and East Central	Narrinyeri	kranti	wullaki	kungari
	Parnkalla	yangkunnu	irallu, yaralta	korti
	Adelaide	kurraki	tiwa	kudlyo
	Darling	kollybooka	pinnya-kollyja	yoongolee
	Diyeri	kudrungoo	—	kootie
	Murunuda	murumiri	kuitch	—
	Mythergoody	karambodla	leepar	—
North Australia and Central Australia	Larrikeya	nangarangwarra (C.)	—	—
	Woolna	lunginmununger	laamal	—
	Daktyerat	mangur	milkbir	dyur
	Ruby Ck., Kimberley	—	—	—
	Napier Range, Kimberley	—	—	—
	Sunday Island	—	—	—
	Macdonnell Ranges	aruilkara	iranta	—
	Walsh R., Q.	—	—	—
	Bloomfield Valley, Q.	pirmbar	kourmabiner	—
	Palmer R., Q.	enbogunby	—	—
	Coen R., Q.	—	—	—
	Mapoon R., Q.	yotte	poonjoo (?)	—
	Gudang, C. York	aira	—	—
Torres Strait	Kowrarega, Torres St.	weama	—	—
	Saibai I., N. G.	—	—	—
New Hebrides	Aulua Malikula	—	—	—
	Nguna	—	—	—
	Aniwa	—	—	—

BIRD	FLY	MOSQUITO	CRAYFISH
puggunyenna	mongana	—	—
punna	monga	redpa (J.)	—
—	—	—	—
iola	—	—	—
mouta mouta (E)	niarnar (N.)	mokerer (N.)	—
kuiap kuiap	—	goguk	talakborong
—	dyodyot	ngoiya ngoiya	bambam
yarboga	bityik	lere	yaabitch
—	bityik	lire	yapitch
—	korgorark	—	yinangi
wotipir	pittik	liriu	nark (?)
—	minik	—	yaam
—	minik	krukrik	weechuk
tuman tuman	ulul	moon-o-erp	konkro, keler
yarri yarri	yilongoure	mundi	—
—	bian	ñuan	dendong
—	maiangamba	—	tongamhalanga
—			naingan
dibbin	burrimal	muggen	—
—	myanga (H.)	doora (H.)	—
tibbin	wumenkan (*pl.*)	toping	—
tigbara	burulu (*pl.*)	mungin	—
dhippi	dhippi, debingo	bunba	illai
widgywidgy	mugguing	buurn	mamuru
jerdie	—	—	gonak
—	kuragura	nuni	—
—	—	—	—
—	am-monga	gee-winnia	—
pulyeri (*small*)	tyilyi	muruli	meauki
irta	yumbarra	yuwtinyu	—
—	tappa	—	—
—	wingoroo, mokay	koondee	koon-goolvo
piya (*pl.*)	moonchoo	koontie	kuniekundi
milkipulo	nango	—	mulpo
—	millua	woonjoin	beekodl
maddo	—	lamda (C.)	—
—	longita	monarongara	—
balbalma	ngätyun	wengnun	—
—	—	—	—
—	—	—	—
—			—
—	ngurur	irol	—
debadeba (*pl.*)	manga (*pl.*)	—	ityanma
—	aigir	—	—
tchekal	kalerwoory	kumu	warkoju
—	amin	ombolum	omothoo
—	troroo	ngoroo	—
—	adhetanne	ngoroo	—
wuroi	wampa	uma	lang-gunya
ure	buli	iwi	kayer
uroi	buli	iwi	kaiaru
nemin	nelang	tongas	—
manu	lango	namu	—
ta manu	arago, anono	tanamu	touretshi

GROUPS	DIALECTS	WORM	SNAKE	ALIVE
Tasmania	*East*	—	loiena	—
	South	—	loinah	—
	West and N. W.	—	rounna rawannah	—
	North	—	powranna (J.)	—
	Miscellaneous	—	katal (L.)	—
Victorian Region	*Yarra R.*	dhuro	kolornung (*black*)	murundak
	Lal Lal	bilitch	kurnmil	muron
	Ercildoune	kurk	kunmil	moron
	Avoca R.	tyumbilitch	kurnmil	muron
	Broken R.	—	kurnmil (*whip*) kulor-nong (*black*)	moronda
	Gunbower	tyumbilitch	kurnwil	murun
	Warrnambool	—	guram	—
	Mortlake	krook	korung	pondean
	Booandik, S. A.	—	—	ur-lea
	Lower Lachlan and Murrumbidgee	tungali	karni	poorwoki
	Gippsland	wurwot	dhurung	beakwan
N. S. Wales and S. Queensland	*Barwidgee, Upper Murray*	—	murray jooyu	—
	Woorajery Tribe, Upper Murray	—	kollojuna (*black*)	—
	Wiraidhuri	duronggargar	during	murron
	Turuwul, Port Jackson		cahn (C.)	—
	Awabakal	—	maiya	moron
	Kamilroi	—	nurai (*black*)	moron
	Kabi, Mary R., Q.	kularen, mune	murang, wongai	manngur
	Warrego R., Q.	gowwa	kan	kurrin
W. Australia and West Central	*Toodyay (Newcastle)*	—	wakal	—
	Pidong	—	millura	—
	Minning	—	mulawuda	—
	Lake Amadeus	—	wom-mee	—
South of S. Australia and East Central	*Narrinyeri*	tyllye, miningkar	kraiyi	tumbe
	Parnkalla	—	nurru (*black*)	ipi
	Adelaide	—	—	purunna
	Darling	—	meetindy, dahngoo	—
	Diyeri	—	woma	thipie
	Murunuda	—	diramatchi	kuli
	Mythergoody	booralkar	chinur	karlir
North Australia and Central Australia	*Larrikeya*	—	midjeera (C.)	medip
	Woolna	—	lermalyer	—
	Daktyerat	dityaruk	ngundyul	karalla (?)
	Ruby Ck., Kimberley	—	—	—
	Napier Range, Kimberley	—	—	—
	Sunday Island	—	murodh, thuro	—
	Macdonnell Ranges	tyaba (*pl.*)	up-moa (W.)	etata
	Walsh R., Q.	—	walkui	aber
	Bloomfield Valley, Q.	ulngermo	tcharper	dorango
	Palmer R., Q.	—	oloor	—
	Coen R., Q.	—	agoye	jeroome
	Mapoon R., Q.	—	agearri	loimre
	Gudang, C. York	—	kanurra (*brown*)	anading
Torres Strait	*Kowrarega, Torres St.*	kurtur	karomat	danaleg
	Saibai I., N. G.	—	tabu	—
New Hebrides	*Aulua Malikula*	—	nemat	maur
	Nguna	asulati	ngmata	mauri
	Aniwa	tanufakere	agata	mouri

DEAD	BIG	SMALL	LONG
mientung bourrack	teeunna, pawpela	canara (J.)	rogoteleebana
moye	papla, proinn, nughabah	teeboack (J.)	rotuli
—	—	—	—
			—
			—
kragbaga	elpenia	bodenevoued	
wordiock (N.)	marrinook (D.)	teebrack (D.)	
wlagaidh	buluto	waiyibo	ñiririmda
dita'a	didibil	nganyagurk	ñirirm
ditai	dyangadya	bupuok	dyowang
ditaiyang	motyauk	wortipuk	tyuarng
werigai	wurdhau	waikurkurong	yorbortak
wadhyingdbya	kurumbirt	martuk	karwil
kolpirna	porir	kunye	wurombit
kalpari	meearong	goomoomoneek	wroroombit
nooan, woora	woorong	moo-ro-ke	woorongbool-e
berapil	kraii	biabi	tiangi
trartigon	kwerail	dhalitch	wragilman
—	murando	bunyungahai	keenyaro
—	kubborn (?)	pooparjol	kubborn (?)
ballun	babbir, burdon	bubbai	bari
boe (H.)	murray (C.)	murruwulung (R.)	kaiun (*large*, R.)
tetti	kauwul	mitti, warea	—
balun	burul	kai, butf	gurar
balun	wingwur	dhomarami	guran
ballyab	darda	gidju	bunderra
winading	goombar	batain	welyardy
nga'arri	yannda	thunthammada	muttharra
—	birdinn	—	—
ill-loong	bun-tor	wee-ma	wat-tora
pornir	grauwi	muralappi	yulde
kunyu, kupa	bumba, mirru	perru	malka
—	parto, witte	kutyo	towinna
bookka	koombaja	kelchelko	berlooroo (*tall*)
narrie	marpoo, pina	waka	—
palpinda	nurda	waga	puri
mujanoo	murdoo	churloo	ooramin
belinying	kuillege	mulutjil (C.)	—
moama	meeania	mee-etniea	—
muruneka	yidello	yigbelderang	tyalala
—	—	—	—
kurdiman, nuniii	weedi	—	—
—	—	—	—
ill-looka	knira	kurka	tanya
lon	molkar	nyolb	wungal
warlan	tchere	boorpan	kalpe
oolbin	—	—	—
mooute	woitte	pfoimakonne	danagoome
moonte	pari	foppari	danagootte
etora, etolma	intonya	embowa	—
uma	keinga	muggingh	kulalle
umanga	kaiza, kai	magina	kukutaligna
emis	lumbon	kakas	barimbarap
mate	warua	kiki	varau
mate	sore	sisi	palo

GROUPS	DIALECTS	SHORT	GOOD	BAD
Tasmania	East	—	elangoonya	noweiack
	South	—	miree	noile
	West and N. W.	—	noangate	ee-ayng-la-leah
	North	—	—	
	Miscellaneous	—	narrarcooper (N.)	katea (L.)
Victorian Region	Yarra R.	morda	manamidh	ñulam
	Lal Lal	murt	wen-gyur	ñulam
	Ercildoune	bunyindyok	dhalkok	yatyang
	Avoca R.	murt	dalkuk	yatyang
	Broken R.	waikorong	burndap	ñulamdak
	Gunbower	tuluandok	dhalguk	yataanduk
	Warrnambool	mulobit	ngutyung	ngamegalin
	Mortlake	mambit	moidhung	amikullien
	Booandik, S. A.	mooter	murtong	wrang
	Lower Lachlan and Murrumbidgee	toonathaigi	primalia	booki
N. S. Wales and S. Queensland	Gippsland	tukalaban	lean, leanman	muratch
	Barwidgee, Upper Murray	koblo	budjeri (?)	—
	Woorajery Tribe, Upper Murray, N.S.W.	nerangi (?)	murrambung	—
	Wiraidhuri	bungulgat	marang	bainguang
	Turuwul, Port Jackson	toomurro (C.)	boodjerre (C.	were (H.)
	Awabakal	—	murrorong	yarakai (to be)
	Kamilroi	bunggudul	murruba	kagil
	Kabi, Mary R., Q.	dhalbur	kalangur	warang
	Warrego R., Q.	wordda	murring	yourral
W. Australia and West Central	Toodyay (Newcastle)	korrat	kwabba	wendang
	Pidong	binnbi	bunndi	wolyi
	Minning	—	puddja	kanung
	Lake Amadeus	—	—	—
South of S. Australia and East Central	Narrinyeri	kopetikke	nunkeri	wirrangi
	Parnkalla	burtu, kartu	marniti, yuwa	milla, nangka
	Adelaide	kurlto	marni	wadli, wakkinna
	Darling	kardooka	gunjulla	toollaka
	Diyeri	wordoo	oomoo	mudlaunchie
	Murunuda	pula	patchi	dira
	Mythergoody	thamin	margul	marthy
North Australia and Central Australia	Larrikeya	—	petyi	goarra
	Woolna	—	mudla	kowarra
	Daktyerat	yindyarok	yunbain	yinetto
	Ruby Ck., Kimberley	—	—	—
	Napier Range, Kimberley	—	—	—
	Sunday Island	—	—	—
	Macdonnell Ranges	botera, dotadota	mara	kuna, mbala
	Walsh R., Q.	—	—	nyumaik
	Bloomfield Valley, Q.	koolger	kombir	booyoun
	Palmer R., Q.	—	oonge	inthe
	Coen R., Q.	beroo	trango	tschooyitanne
	Mapoon R., Q.	beroo	peroo	niattapfreune
	Gudang, C. York	—	—	
Torres Strait	Kowrarega, Torres St.	—	kape	wate
	Saibai I., N. G.	taupainga	kapu	wati
New Hebrides	Aulua Malikula	burunk	embu	esamp, umwi
	Nguna	vuru	wia	sa
	Aniwa	poto	erifia	isa

HUNGRY	THIRSTY	RED	WHITE
—	rukannaroonyack	tentya	malleetye
—	kukannaroitee	koka	mallee
—	—	—	mungyanghgarrab
—	—	bolouine	lore
plonerpurtick (N.)		—	
ñiraburdinan	konboningan	bïpïdharnin	dharanun
miraiauwirmo	kurtnongin	dirkwarin	tararapil
milaia	koönma	ñurong	dardanitch
mi-laiang	kuunmon	bitudyan	tardarnit
ñirebirnang	konbuninyan	dïrbarïn	dhirarañun
wïkanda	borgunyinda	nerwail	dhorathauil
bardubangulanga		kirikirigunitch	
barda-bong-wothone	kookuongbaritha	batkoitch	apkooitch
dritban	koornonine	kro-milit	marmon
kraibira	koornoman	kooroorgandi	pliandı
kanyugon	kuan-guran	kurrgirik	dhabon
bungunowo	jargenauer	—	—
—	wijela	girri girri	—
ngarran, yuar	ngandabirra	diren direng	barrang, ngalar
yuroo (C.)	—	morjal (H.)	taboa (H.)
kapirri (*to be*)	—	tirriki	—
—	kollengin	koimburra, gue	pullar
gandho	yallo, ngaiallo	bothar, kuthing	kakal
koundal	biruboliu	murgy murgy	buada
ulup	gabby ulup	noba	—
nyourru	minni	billyini	billon
—	nan-too	olba (*red ochre*)	lill-lill
yeyauwi	klallin	kurungulun	balpi
mai-karnba	yernpiti	—	palkara
taityo	—	karro karro	perkanna
wilkahka	yerlkka	nahllkeeka	bichooka
mooalle	murdiealie	murulyie	booloo
munkuwaninga	napawapinanie	katachuka	—
pulningoo	urbingoo	cilcilgarco	boonaroo
amanding (*I am*)	golapping	—	arnarra
unggwerdea	immocaia	merwaler	lunginmunnunger
manorik	puin	witma	tamarma
—		—	
marigan	burra	—	toop, milli
—	—	—	—
—	ankatala	tataka	—
ongguair	honggir	aiguir	—
tchakoi	bannerga wahou warli	marun	pingaji
ange	ingky	—	—
—	ngoitschi	arrumbre	—
adhaimre	tedikka	arrumbre	arroa
awora	—	—	—
weragi	nuk enei	kulkthung	uru
mitniginga (*hunger*)	—	kurkagamulnö	ejamulnga
merangaskas	minrok	miel	embusa
pitolo	maru	miala	tare
tshitage	mate tovai	ouri, ouroura	kego

Groups	Dialects	Black	Full	Empty
Tasmania	*East*	mabanna	rueeleetipla	—
	South	loaparte	kanna	—
	West and N. W.	—	yeackanara	—
	North	wadene wine	—	—
	Miscellaneous	—	—	—
Victorian Region	*Yarra R.*	wurgarin	tyavbun	dhindivik
	Lal Lal	wurkarapil	yirt'no	ngamgarin
	Ercildoune	worwoganitch	tirnda	—
	Avoca R.	wurukutyauil	purntya	dakerang
	Broken R.	wurgabil	dyabuin	kalarmun
	Gunbower	kunetyulauil	tyindilauil	dhulkain
	Warrnambool	miln	—	—
	Mortlake	meeing	memug bakka	bangathung
	Booandik, S. A.	woorlo	—	—
	Lower Lachlan and Murrumbidgee	waikerimbi	wonounna	terawna
	Gippsland	ninbon	tandurgon	tenyugon
N. S. Wales and S. Queensland	*Barwidgee, Upper Murray*	—	—	—
	Woorajery Tribe. Upper Murray, N.S.W.	—	—	—
	Wiraidhuri	buggabugga	urrur	bain bain
	Turuwul, Port Jackson	nand (H.)	boruck (C.)	parrat berri (C.)
	Awabakal	puto (*to be b.*)	—	—
	Kamilroi	bului	yularai	—
	Kabi, Mary R., Q.	mullu	gumka	nolla
	Warrego R., Q.	kurda	dadh-biru	—
W. Australia and West Central	*Toodyay (Newcastle)*	moon	—	—
	Pidong	wiri	—	—
	Minning	—	—	—
	Lake Amadeus	mar-roo	—	—
South of S. Australia and East Central	*Narrinyeri*	kineman	yalkin	pek
	Parnkalla	mau-urru	bakkamba	karnba
	Adelaide	pulyonna	buttonendi	—
	Darling	yerrelko	—	—
	Diyeri	muroo	—	—
	Murunuda	—	waponurda	wapowagina
	Mythergoody	margin	walgillbongo	bulninyu
North Australia and Central Australia	*Larrikeya*	binyuminnkoe (P.)	gager	kwaotidong (P.)
	Woolna	—	—	—
	Daktyerat	eyukeyuk	arugunuka	pinyuya
	Ruby CK., Kimberley	—	—	—
	Napier Range, Kimberley	—	manar	—
	Sunday Island	—	—	—
	Macdonnell Ranges	urbula	—	—
	Walsh R., Q.	—	arbun	—
	Bloomfield Valley, Q.	ngombo	tchakal	yamberkari
	Palmer R., Q.	—	—	—
	Coen R., Q.	nambarra	angakapaddi	tschoramme
	Mapoon R., Q.	nambarra, 'mbre	angapit	arramme
	Gudang, C. York	—	—	—
Torres Strait	*Kowrarega, Torres St.*	kubi-kubi thung	—	—
	Saibai I., N. G.	kubikubinga	—	—
New Hebrides	*Aulua Malikula*	miet	embura	nesungun
	Nguna	loa	vura	kpalo
	Aniwa	uri	fonu	noaga ana

Quick	Slow	Blind	Deaf
—	—	—	guallengatick
—	—	—	guanghaia
—	—	—	wayeebede
—	—	—	—
—	—	—	—
yarbok	bainggongak	turtmirng	turtwirng
wariwi	bulkal kulne	nyima	bong bong
mondap	burtai	ñim ñim	nga nga
wariwi	bulkal kulne	ñima	nga nga
yuarbok	baingongak	borm borm mirng	nga nga wirng
werkuk	paatoka	bormail	dhapilaurimbul
—	—	kunditch	ngurdinwin
—	bangaatong	krooncheebur	moorkin wirn
—	—	kolo porn	netingwrungung
minanaw	yalimongi	panmapll	markenki
wedhur	yardoman	murindan	naringon
wungurela	—	megeewanjega	megee murrumbugga
burrabari	woori burrabari	mookeer	megootha
nanan	—	ngamabang	ngia mugga
—	—	—	—
—	—	—	—
kaiabur	bullo, malo	muga	mugabinna
kalu, dhalli	yul	mi-gulum	pinang gulum
kurdin	wulling	nurnding	mugubinna
yatta golly	dabicin (?)	gennang wadder	dwangoburt
—	—	—	—
—	—	—	i-rita
—	mant	tonde	plombatyc
narru narru	widlara	mena wapo	yurre ngundanniti
maityukka	mantikatpa	padyotti	—
—	—	wontooja	nahppaja, moko
nooroo	—	bootchoo	kootcharabooroo
munkaobi	munkwapi	puitchi	pingatuda
bodlun	niju niju	waramugu	pennekalunu
kuillibik	—	dlamon-ngapinga (P.)	kwaella buellyidong (P.)
merpur	dama	woin	ngamama
—	—	—	—
—	—	—	—
—	—	—	—
ilkaunkuanta	—	—	i-rita (W.)
adbel	yenda	libwon	piarath
yeakere	yambal	boorer	milger kari
—	—	—	—
—	—	andamu	woititre
tre	bwoi	andamu	wumreschatti
—	—	—	—
—	—	—	wate kowrare
—	—	bugiri	moamai kuralna
biltah turur	—	metina embara	wisina boiimboro
ngmaravarave	aliali	kesa	kparo
miloulou	fakasisi	kofu foimata pouri	tuturu

Groups	Dialects	Strong	Weak	Heavy
Tasmania	East	oyngteratta	koomyenna	miemooatick
	South	rulla rullanah	mia wayleh	moorah
	West and N.W.	ramanarrale	—	—
	North	—	—	—
	Miscellaneous .	noormeanner (N.)	—	—
Victorian Region	Yarra R. . .	bonmarart	yaralornin	barnborn
	Lal Lal . . .	balert	palka ba-ngik	banbon
	Ercildoune . .	martuk	dermderm	bunurt
	Avoca R. . .	punurt	kunamilangan	bunurt
	Broken R. . .	borndop	—	barndarbun
	Gunbower . .	tormoil	tyipor	kunkimilong
	Warrnambool .	—	—	kurgirnitch
	Mortlake . . .	arakitchung	wahmpehur	parpurnaman
	Booandik, S. A.	—	—	koon goon
	Lower Lachlan and Murrum-bidgee	wongorapi	wongathe	wenthepil
	Gippsland . .	tardiman	—	kurkuran
	Barwidgee, Up-per Murray .	matong	mulumbudji	boobobelo
N. S. Wales and S. Queensland	Woorajery Tribe, Upper Murray, N.S.W.	metong	woori metong	—
	Wiraidhuri .	ginnar, wallan	gamban	maddo
	Turuwul, Port Jackson	—	—	—
	Awabakal . .	—	—	porol (to be h.)
	Kamilroi . .	warunggul	—	munan
	Kabi, Mary R..Q.	bauguthar	naman mokkan	dbikir
	Warrego R., Q.	murrabinbirra	wallabinbirra	domming
W. Aus-tralia and West Central	Toodyay (New-castle)	murdich	—	—
	Pidong . . .	—	—	—
	Minning. . .	—	—	—
	Lake Amadeus .	—	—	poonta buckanee
South of S. Aus-tralia and East Central	Narrinyeri . .	piltengi	pultne	talin
	Parnkalla . .	ngalliti	kappara	yukarta
	Adelaide. . .	taingiwilla	mannanya	yuruti
	Darling . . .	koorkree	eella-koorkree	—
	Diyeri . . .	—	—	murdie
	Murunuda . .	peuri	punchira	nurda
	Mythergoody .	nowargoodul	nowarkulunga	oolmul
North Australia and Central Australia	Larrikeya . .	kuillege, dangkal	—	mutki
	Woolna . . .	lerwinyueker	—	—
	Daktyerat . .	yingnelek	nganburk	turma
	Ruby Ck., Kim-berley	—	—	—
	Napier Range, Kimberley	—	—	—
	Sunday Island .	—	—	—
	Macdonnell Ranges	ekalta, ntatna	taltya	inbora, yotia
	Walsh R., Q. .	—	—	mulkar
	Bloomfield Val-ley, Q.	ngaraji	ngara kari	kolongul
	Palmer R., Q. .	—	—	—
	Coen R., Q. . .	patowoittrekke	patotea	angonoo
	Mapoon R., Q. .	pfui woittrekke	pfui tea	angonoo
	Gudang, C. York	—	—	—
Torres Strait	Kowrarega, Torres St.	—	—	mapule
	Saibai I., N.G.	—	—	—
New Hebri-des	Aulua Malikula	bahario	se bahario	merans
	Nguna . . .	kasua	manainai	maranga
	Aniwa . . .	tomatua	taru	mafa

Light (Not Heavy)	Afraid	Sweet	Right
—	tianna coithyack	—	—
—	tiennawille	—	—
—	camballetc	—	—
—	—	—	—
—	—	—	—
bular-ornin	pambun	kiringkirm	ngaiabunburndap
wulrung	ngalblinyan	kepgip	waingur
daap	baamba	giagia	nardodalodye
tap	bambun	kepgip	tatkuk
wulurndyak	bambun	—	—
mormor	paamba	wityer	dhalkungok
dhalap	kurninba	—	—
thalup	coninbanon	woombool	oochoag
tap	yinoon	—	murtonga
naimno	kaingon	primalia	primalia
bauugan	dhiragon	leanmon	lean
baumbaji	—	—	—
—	—	—	—
ginarginar	giarra (*to be*)	ngarrungarra	marrombul
—	bargat (C.)	—	—
—	kuita (*to be a.*)	—	—
—	gial, ghilgbil	kuppa	—
nandimathi	widhiman	geyar	yamba
walladomming	kurra	wian-kulla	birndal
—	wain	nil	kwaba
—	—	—	—
—	—	—	—
—	—	—	—
kaikai	blukkun	kinpin	nunkeri
yalluru	waiinniti	ngaltya, nganyara	nalka
baltarta	waiwai	—	numma
—	oollya	gunjulla	—
—	yaupunie	alkooelie	—
waga	kinindu	windra	patchi
barple	kowinjar	churkulingu	mugle
—	—	dadbungua	—
—	nginmar	warkie	—
ngalwar	elindyur	wi	—
—	—	—	—
—	—	—	—
—	—	—	—
—	itnora	unkuala	numbaka
nyolba	lim	—	—
boortal	yinil	kukkan kukkan	ngoulkoor
—	—	inboo	—
tschora	adhete	rollamme	parlimmi
tschora	adbete	rollamme	parliminrimminne
—	—	—	—
—	—	—	—
—	—	—	—
memer	metah	garahar	pangase
masalesale	mataku	mami	leana
mama	kumtacu	mugaro	erefia

GROUPS	DIALECTS	WRONG	STRAIGHT	CROOKED
Tasmania	*East*	miengana	uŋgoyeleebana	powena J.)
	South	nuyeko	tungbabe	—
	West and N. W.	—	—	—
	North	—	—	—
	Miscellaneous	—		—
Victorian Region	*Yarra R.*	ballirt	yurtin	nugim nugimdyirin
	Lal Lal	ñulam	ñirirm	nguring nguring
	Ercildoune	yatya	yulp	ngoningoning
	Avoca R.	yatyang	yulop	nguning nguning
	Broken R.			
	Gunbower	nangutan	yulɔ	widhidyirang
	Warrnambool			
	Mortlake	arnrigullen	dhaarn	wurt wurt
	Booandik, S. A.	wrang	—	weriner
	Lower Lachlan and Murrumbidgee	warta primalia	uiethe	toorapil
	Gippsland	denbon	tutburutbon	wali wali
	Barwidgee, Upper Murray	—	—	—
	Woorajery Tribe, Upper Murray, N.S.W.	—		—
N. S. Wales and S. Queensland	*Wiraidhuri*	wammang	dulluwarai	dalgang
	Turuwul, Port Jackson	—	—	—
	Awabakal	—	—	warin warin (*to be*)
	Kamilroi	—	waragil, gura	—
	Kabi, Mary R.,Q.	waa, warang	dhurun	warkun
	Warrego R., Q.	ural	bindal	worroungouring
W. Australia and West Central	*Toodyay (Newcastle)*	windang	—	—
	Pidong	—		—
	Minning	—		—
	Lake Amadeus	—	chu-cowra	que-ar
South of S. Australia and East Central	*Narrinyeri*	wirrangi	thure	kulkuldi
	Parnkalla	nanna, wadli	inba, yau-urru	ngurdli
	Adelaide	wadli	madurta	yokunna
	Darling	—	—	—
	Diyeri	chika	thalkoo	koontiekoontie
	Murunuḑa	tira	patchi	tira
	Mythergoody	waraburnu	toortoojoo	kungul
North Australia and Central Australia	*Larrikeya*	—	kuinyaki	gurnamadinga (P.)
	Woolna	—	—	—
	Daktyerat	—	dur	gurrurkgururk
	Ruby Ck., Kimberley	—	—	—
	Napier Range, Kimberley	—	—	—
	Sunday Island	—	—	—
	Macdonnell Ranges	bala, mbala	aratya	—
	Walsh R., Q.	—	—	—
	Bloomfield Valley, Q.	buyoun	tchoonke	kuru kuru
	Palmer R., Q.	—	—	—
	Coen R., Q.	anaittakke	brammanjinne	lotroo
	Mapoon R., Q.	anaittakke	brammanjinne	loti
	Gudang, C. York	—	—	—
Torres Strait	*Kowrarega, Torres St.*	—	—	—
	Saibai I., N. G.	—	—	—
New Hebrides	*Aulua Malikula*	esamp	mentement	kambakambul
	Nguna	trangele	leana	tragele
	Aniwa	isa	totonu	safi

Wood Spear	Reed Spear	Shield	Tomahawk
perenna	—	—	—
pe-na	—	—	—
pana, pilbah	—	—	—
racca (J.)	—	—	—
arlenar (N.)	—	—	—
dhar	dhirar	kiarm	morang
karp	tyark	molka	kalpalingork
darbokarok	dyark	malka	bartik
kuyun	tyark	molka	bortik
kolyon	dyeror	dyirom	gargen
kuiyun	tyirom	molga	dhir
ter	terkun	malk	muitdyir
narlmul	dhuruk	malkar	par par koort
kooen	—	malkar, brooal	karkobe
noodlii	ponondi	kuaikuli	tiennie
buran	kauat	bamork	kuean
jerrambahai	yaarga	birregambo, murga	moodewa
—	jeereel	murka	thowan
dullu	—	marga	burguin
kamai (H.)	—	elimang (H.)	mogo (H.)
worai	—	koreil	pukko
pilar	—	bumai burin	yundu
konni	—	kunmarim	muyum
kullia	—	burgu	dowing
gigie	—	wonta	koddue
kulbadi, kurada	—	unda	yarrawa
—	—	—	—
cadgee	—	coor-tichie	—
wundi	kaiki	wakkaldi	drekurmi
kaya, winna	—	—	kandi
kaya, winda	—	—	—
kulkaroo	—	woollombarra	wokkaka
kulthie	—	pirrauma	—
windra	—	kumbura	—
koonjul	—	goonbarra	maree
malagirma	tjinbala (C.)	—	marangima (C.)
mowowie	lilcorla	—	larlinganda
daruk	mokalin	—	litpurp
kurna	—	—	—
bilara	—	—	—
—	—	—	—
tyata	—	al-quirta (W.)	—
yirr	—	pijerikan	wainmil
kalka	—	koongeri	ti-i
ulka	koongoon	koolmurra	egan
—	—	agai	—
woitte	angame	mennti	adhcagge
alka	—	—	aga (*are*)
kalaka	—	—	aga (*are*)
kubai	—	bada	goba
sare	—	—	sip sip, tangata
naio	nalua	—	tangoto
tatou	tatou tagosau	—	toci fatu

Groups	Dialects	Stone Knife	Boomerang	Club
Tasmania	East	—	—	—
	South	—	—	lerga (D.)
	West and N. W.	—	—	—
	North	—	—	rocah (J.)
	Miscellaneous	teeroona (S.)	—	lillar (N,)
Victorian Region	Yarra R.	mung (*quartz*)	wankim	molka
	Lal Lal	gurin	wan-gim	warawar
	Ercildoune	—	derm derm	leawil
	Avoca R.	petch	datom datom	birpin
	Broken R.	kalburn kalburnin	wangim	kudyuron
	Gunbower	dyark	wan	pirbang
	Warrnambool	—	ledam ledam	malingannk
	Mortlake	morijir	lata latup	warawara
	Booandik, S. A.	—	ketum ketum	booamba, kana
	Lower Lachlan and Murrumbidgee	kalingali	onei	moonopi
	Gippsland	dheradherak	wangin	kalak, donmong
N. S. Wales and S. Queensland	Barwidgee, Upper Murray	—	wongewa	goojuroo
	Woorajery Tribe, Upper Murray, N.S.W.	—	bergan	—
	Wiraidhuri	guingal	bargan	girang
	Turuwul, Port Jackson	—	—	ngallung ulla (C.)
	Awabakal	kullingtiella	turrama	kottirra
	Kamilroi	—	burran, burrigul	murula pundi
	Kabi, Mary R..Q.	dhakke	boran	kuthar
	Warrego R., Q.	—	murley	bundy
W. Australia and West Central	Toodyay (Newcastle)	dahba	kilie	dowak
	Pidong	—	wolanu	—
	Minning	—	—	—
	Lake Amadeus	wom-ba (*knife*)	—	wonna (*stick*)
South of S. Australia and East Central	Narrinyeri	drekurmi	panketyi	plongge, kanake
	Parnkalla	yakko	wadna	katta
	Adelaide	yakko		katta
	Darling	yernda	wonna	poondee, koloroo
	Diyeri	yootchoowonda	kirra	—
	Murunuda	kalcichipera	tira	puninanga
	Mythergoody	mindee	kalkarbooey	miraloo
North Australia and Central Australia	Larrikeya	maramari (P. *knife*)	—	—
	Woolna	—	—	metpadinger
	Daktyerat	malauer		tyantan
	Ruby Ck., Kimberley	—	kurli	—
	Napier Range, Kimberley	—	kaili	—
	Sunday Island	—	—	—
	Macdonnell Ranges	irgalla	ulbainya	—
	Walsh R., Q.	—	ongwol	wur
	Bloomfield Valley, Q.	biniu	wangi	doure
	Palmer R., Q.	—	mulkarra	—
	Coen R., Q.	—	—	—
	Mapoon R., Q.	—	—	—
	Gudang, C. York	—	—	ekara
Torres Strait	Kowrarega, Torres St.	—	—	kobai
	Saibai I., N. G.	giturika (*knife*)	—	—
New Hebrides	Aulua Malikula	—	—	menriki, numbot
	Nguna	—	—	nakpe
	Aniwa	tomatshira fatu	—	tererakou

HEAD	HAIR	EYE	FACE
oolumpta	poinglyenna	mongtena	niengheta
poiete	poiete longwinne	nubre, nubrenah	noienenah
eloura	kide	pollatoola (J.)	maurable (J.)
ewucka (J.)	parba (J.)	elpina	—
cuegi (P.)	lagurnar barnar (N.)	neurikeenar (N.)	niperina (L.)
kauwong	yara kauwong	mirng	mirng-bang
moork	ngarmurk	mi	mirbang
burp	rimbil	mir	mirbaga
burp	ngara	mi	mirpaga
kawang	yirikawang	mirng	minyinbokangin
muranyuk	ngoranyuk	milnuk	milnuk
pim	ngarat	mir	mir
beam	wirin	meringh	methin
boop	ngoorla	mir	—
poapi	nouobopi	meingi	biingi
buruk	ledh	mirri	kung
murriawa	murriawa	wunjubba	wahroa
bollong	—	mil	—
kuppura, ballang	uran	mil	ngolong
caberra (H.)	diwarra (H.)	mi (H.)	—
wollung, kuppura	kittung	ngaikung	—
gba, kaoga	tegul	mil	—
kam	dbilla, bon	mi	ngu, wanggum
bambu	worlba	meel	nunga
katta	katta	meal	wonnul
mugga	kulawil	kuru ($pl.$)	—
kada	wenndu	meyl	—
cutta	hoo-ray	coo-roo	—
kurli	kuri	pili	petye
kakka	butti kurni	mena, mialla	mangu, ngarri
makarta	padlo, yoka	mena	murki
turtoo	turtoo-woolkky	meeky	—
mongathanda	para	milki ($pl.$)	—
kuncbo	nircha	milki	mula
kundra	jamul	euko	—
maloma	balrrijeen (C.)	damörra	darreminndbirra
mudlo	imalgnie	ma	—
pondo	pondomer	numuru	bebema
ungunyangunya	lungga	milwa	—
mandin	lamingar	nimilar	—
nalma	numandadi	nimi	—
kaputa	gola	alkna	angnera
harui	wir	lipwar	je
tokol	monger	mil	wallau
ambogo	allung	immun	—
drokke	ea	andoa	woikapoo
ranrui	ranrii	andoa	waggapoo
pada	odye	dana	—
quiku	yal	dana	—
kuiko	yalbupu	purka	paru
batina	nepol batina	metina	konin
nakpauna	naluluna	namatana	namatana
nouru	toura	foimata	foimata

Groups	Dialects	Ear	Nose	Smell (Noun)
Tasmania	East	mungenna	mununa	—
	South	wayee	muye, muggenah	—
	West and N. W.	lewlina (J.)	muanoigh	—
	North	tiberatie	medouer	—
	Miscellaneous	wegge (R.)	mudena (R.)	—
Victorian Region	Yarra R.	wirng	kang	buang
	Lal Lal	wirng	kang	bang
	Ercildoune	wirmbol	ka	ngarba
	Avoca R.	wirbul	kra	buwang
	Broken R.	wiring	kangin	buang
	Gunbower	wirmbuluk	kañuk	buanga
	Warrnambool	wirn	kapung	wapirna
	Mortlake	methin	kabo	woomban
	Booandik, S. A.	wiung	kow	—
	Lower Lachlan and Murrumbidgee	maarki	tiendi	naarota
	Gippsland	wuring	kung	meabilon
	Barwidgee, Upper Murray,	murrumbo	dendewa	—
N. S. Wales and S. Queensland	Woorajery Tribe, Upper Murray, N.S.W.	wootha	merootha	—
	Wiraidhuri	uda	murru	buddur buddur
	Turuwul, Port Jackson	gorai (H.)	nogur (H.)	—
	Awabakal	ngureung	nukoro	—
	Kamilroi	binna	muru	—
	Kabi, MaryR., Q.	pinang	muru	ka
	Warrego R., Q.	binna	nuru	buddley
South of S. Australia and East Central / W. Australia and West Central	Toodyay (Newcastle)	dwanga	moolye	—
	Pidong	kulka (pl.)	mulya	—
	Minning	gula	mula	—
	Lake Amadeus	pinna	am-mou-la	—
	Narrinyeri	plombi	kopi	—
	Parnkalla	yurre	mudla	kurbo
	Adelaide	yurre	mudla	marto
	Darling	eurree	pulkkapinna	—
	Diyeri	cootchara (pl.)	moodla	koolie
	Murunuda	nura	mula	tunka
	Mythergoody	pinul	yeengar	noomalbunju
North Australia and Central Australia	Larrikeya	banarra (C.)	queeanguar (C.)	—
	Woolna	wal	weer	—
	Daktyerat	monindyaur	yinun	ngeadurkma
	Ruby Ch., Kimberley	munga	dirdi	—
	Napier Range, Kimberley	nilibib	gurmil	—
	Sunday Island	nilimar	nemal	—
	Macdonnell Ranges	ilba	ala	—
	Walsh R., Q.	wi	je	—
	Bloomfield Valley, Q.	milger	pujil	ngoumal
	Palmer R., Q.	innur	omo	—
	Coen R., Q.	woie	kokanne	—
	Mapoon R., Q.	wogo	ri	—
	Gudang, C. York	ewunya	eye	—
Torres Strait	Kowrarega, Torres St.	kowra	piti	ganu
	Saibai I., N. G.	kaura	piti	ganu
New Hebrides	Aulua Malikula	ursina	ngunsenda	nebohte
	Nguna	naraligana	ragisuna	—
	Aniwa	tarega	nousu	namu

Mouth	Lip	Teeth	Chin
kakannina	—	wugherrinna	comnienna
kaneinah	—	pay-ee-a	wahba
kapoughy leah	—	yannalope (J.)	—
mona	mona (*pl.*)	iane	—
canea (J.)	wurlerminner (N.)	leeaner (*sing.* N.)	congeue (R.)
worong	worong	liang	ngorndak
wuru	wuru	liang	ngandak
wurobodhali	wuro	lia	nganyi
wuru	wuru	lia	ngonyi
wurungin	wurungin	liang, lang	—
tarbuk	wuruñuk	liañuk	pumaniñuk
ngulom	wurung	tangam	—
woorong	woorong	dhungun	orine
lo	wro	tunga	—
warongi	belathowongi	naroki	nharlki
kaat	yungaat	ngondok	yien
niwa	—	niyu	lendawa
yabba (?)	—	yeerong	—
ngan	—	irang	yaran
kalga (H.)	willin (*pl.* H.)	dara (H.)	wallo (H.)
kurrurka	tumbirri, willing (*pl.*)	tira	wattun
—	ille, kumai (*pl.*)	yira, ira	tal
dbangka	dambur	dhangka	yikkal
dad	mimmi	thir-ra	yeroghin
—	—	nannang	—
ira	wuti	willga	nganngu
thumminji	—	iri	—
tar	tar-bin-bimba	car-teta	noo-too
tori	munengk	turar	nguiture
ya	nemi	ira	ngangunge
ta, narparta	tamino, tamandi	tia	nguttoworta
yelka	moonnoo	nunndee	wokka
muna	miemie (*pl.*)	munathandra	unkachanda
dira	dira	malilku	nancha
yarcharain	tharingar	yerdidther	yanbar
gurbalquar (C.)	—	unbirregee (C.)	gonngonngwa (P.)
wabn	waper (*pl.*)	ya	—
aru	aru	dir	tdang
lira	—	—	—
yallar	—	—	—
nilyi	—	—	—
a-ruck-a-ta (W.)	arinbinba	deta, detya	rotna
—	jil	wea	artchan
andel	yimbi	tchira	bari
amitting	—	ookool	—
anga	kai	adhetroombao	angao
anga	kago	mapibao	angao
angka	angka	ampo	ebu
guda	iraguda	danga	ibu
gudö	iraguda	dang	gunga
bangona	nakulu bangona	nelvanta	mesembrin
nawokana	nangolena	napatina	qasina
rogoutu	nokiri ragoutu	nonifo	nocuincumi

Groups	Dialects	Cheek	Forehead	Beard
Tasmania	East	—	raoonah	comena purennah
	South	—	roee roeerunna	cowinne
	West and N.W.	—	rioona	comene waggele
	North	neprane	—	kide
	Miscellaneous	nobittaka (D.)	monur (N.)	kongine (P.)
Victorian Region	Yarra R.	wangga	minyin	yaragorndok
	Lal Lal	wang	mīn	ngarangandak
	Ercildoune	murak	gini	nganyibauro
	Avoca R.	murak	kinnī	ngcnyi
	Broken R.	wanggik	minyin	ngornang
	Gunbower	murakuk	kiniñuk	ngarañuk
	Warrnambool	dhakka	mitdhin	nguran
	Mortlake	dhurthuk	nethin	—
	Booandik, S. A.	wuraa	kine	ngurla ngerne
	Lower Lachlan and Murrumbidgee	nurni	kernangi	monangi
	Gippsland	wong	nin	lidh
N. S. Wales and S. Queensland	Barwidgee, Upper Murray	—	—	yarangba
	Woorajery Tribe, Upper Murray, N.S.W.	—	—	—
	Wiraidhuri	daggal	ngolong	yaran
	Turuwul, Port Jackson	—	nulla (H.)	yarrin (C.)
	Awabakal	kullo (*pl.*)	yintirri, ngollo	yarrei
	Kamilroi	—	ngulu	yare
	Kabi, Mary R., Q.	wanggum	nulun	yeran
	Warrego R., Q.	nummung	bubbal	yerreng
IV. Australia and West Central	Toodyay (Newcastle)	—	—	nanga
	Pidong	—	bulla	nganngul
	Minning	—	—	—
	Lake Amadeus	per-till-lerra	null-ar	un-gurra
South of S. Australia and East Central	Narrinyeri	make	bruye	menaki
	Parnkalla	kalba, ngulko	ngarnka	kanbanggurru
	Adelaide	malaitye	murki, yurlo	malta, yernka
	Darling	nullee	beekkoo	wokka woolkky
	Diyeri	—	milperie	unka
	Murunuda	ulcho	pilpa	uncha
	Mythergoody	walindu	themer	mangoora
North Australia and Central Australia	Larrikeya	—	mudpirrma (P.)	gueabalma
	Woolna	—	—	yaba
	Daktyerat	ngaruk	milk	marabat
	Ruby Ck., Kimberley	—	—	nunga
	Napier Range, Kimberley	—	—	alungar
	Sunday Island	—	—	dhird
	Macdonnell Ranges	ilgaia (*pl.*)	litna	ininga
	Walsh R., Q.	gul	halban	alpar
	Bloomfield Valley, Q.	gangool	yiman	wallar
	Palmer R., Q.	—	—	aworko
	Coen R., Q.	ngone	pai	nga
	Mapoon R., Q.	ngone	pai	nga
	Gudang, C. York	baga	eprinya	yeta
Torres Strait	Kowrarega, Torres St.	baga	paru	yeta
	Saibai I., N. G.	—	pautö	yata
New Hebrides	Aulua Malikula	misembinta	—	nepol mesembrin
	Nguna	napupuna	naraena	naluluni nasina
	Aniwa	marigariga	nomugarai	nofurfuri cumu

TONGUE	STOMACH	BREASTS	ARM
kayena	teenah	parugganna	wu'hnna
menue, maynah	teena	parugganah	wu'hnna
tullana (J.)	teenah	—	alree (J.)
guenerouera	maguelena	pouketalagna	anme (*forearm*)
kanewurrar (N.)	plaangner (N.)	wagley (J.)	wornena (R.)
dhalang	hoitch	birm birm	derak
dyilang	poitch	tyuram	torak
tyale	wutyop	gurm	datyak
tali	billi	tyang	tatyak
dyelang	bodyin	—	—
dhaliñuk	wudyumbuk	dyanguk	tatakuk
dhulan	tokung	ngabung	wurk
dhaline	dhogogang	mart	woork
tale	boole	—	woo
turlengi	belangi	koimbi	tarki
dyelan	bulun	bang	brindang
dullingba	—	—	karjenba
—	boorban	—	merrola
dalan	binbin (*belly*)	duddu	—
tallang (C.)	barrong (H.)	nabang (H.)	gading (H)
tullun	purrang (*belly*)	paiyil	—
tulle	mubal (*belly*)	birri, ngummu	bungun
tunam	dungun, kuri	among	kining
durling	duggu	durley	biggey
dalain	gobbil	bibbi	mara
thallin	warri	—	katti murrnun
thalidd	mukulla	—	wanngu
—	weelar	ip-pee	minna
tallanggi	mankuri	ngumpurengk]	tyele (*upper arm*)
yarll	ngangkalla	ngamma	ngando, yurti
tadlanya	ngankimunto	ngammi (*s.*)	turti
tulleenna	koonto	poonna	wunyee, mungko
thulie	mundra koodnabidie	auma	oona (*pl.*)
tali	wapunurda	muna	cilka
tumingaroo	uparer	uminar	waljur
kulamelloa	quallama (C.)	mamabilma (C.)	kwiaverndara (P.)
wee-e	marna	ngolya	leuveyer
ngandork'	mandulma	wing	wuru
—	—	—	—
—	nung	mamini	yarmilar
—	—	—	—
alinya	idunta	ibatyangna (*pl.*)	inanga
are	—	pip	dhom
nabil	wahral	bi bi	tchahil
elpin	oroom	onyong	—
ngai	arra	anjoou	aga
lanne	arra	anjoou	mearri
untara	maita (C.)	yongo	—
nai	wera	susu	—
boia	maitarun	da (*sing.*)	getö
lemen	tamban	nisna	verna
namenana	napoloalapa	nasusuna	naruna
norero	tupewa	afatfata	norima

Groups	Dialects	Hand	Finger	Nail
Tasmania	East	riena	riena	tonye
	South	reemutta	rye-na	ryeetonye
	West and N.W.	—	reeleah	wante leah
	North	rabalga (J.)	anme	nil (*pl.*)
	Miscellaneous	anamana (L.)	beguia (E.)	pereloki (E.)
Victorian Region	Yarra R.	mornang	mornang	dhirip
	Lal Lal	mona	mona	dirip
	Ercildoune	manya	manya	lirn manya
	Avoca R.	monya	monya	lilli
	Broken R.	—	—	—
	Gunbower	manañuk	wotit manañuk	liri manañuk
	Warrnambool	morang	morang	pirn morang
	Mortlake	muruk	muruk	brin muruk
	Booandik, S. A.	murna	murna (*pl.*)	—
	Lower Lachlan and Murrumbidgee	mumangi	naraugori	larimongngi
	Gippsland	brety	dhakirbret	dhakirbret
N. S. Wales and S. Queensland	Barwidgee, Upper Murray	murra	—	—
	Woorajery Tribe, Upper Murray, N.S.W.	murra	—	—
	Wiraidhuri	marra	—	yulla
	Turuwul, Port Jackson	tamira (H.)	berrille (H.)	carrungen (C.)
	Awabakal	mutturra	—	tirri
	Kamilroi	murra	—	yulu
	Kabi, Mary R., Q.	piri	piri, molla	molla, gillen
	Warrego, R., Q.	murra	murra	pigging
W. Australia and West Central	Toodyay (Newcastle)	—	—	—
	Pidong	murra	—	minndi (*pl.*)
	Minning	murra	murra-kabudd (*pl.*)	—
	Lake Amadeus	murra	murra	mill-tee
South of S. Australia and East Central	Narrinyeri	mari	turnar	perar
	Parnkalla	marra	marra	birri
	Adelaide	marra	marra	biri
	Darling	murra	—	mellinya
	Diyeri	murra	murramookoo (*pl.*)	murrapirrie (*pl.*)
	Murunuda	mira	pinga	—
	Mythergoody	mumbinoor	malbidji	malbidji
North Australia and Central Australia	Larrikeya	kuiaroa (C.)	gwiarrwoa (P.)	daalla (P.)
	Woolna	itpaya	tyanamunger	—
	Daktyerat	nanyulk	yinbar	pir
	Ruby Ck., Kimberley	murli	—	mildha
	Napier Range, Kimberley	—	—	—
	Sunday Island	nemala	—	oral
	Macdonnell Ranges	iltya, raga	iltyaganya	itapmara
	Walsh R., Q.	dhi	dhi	—
	Bloomfield Valley, Q.	mara	mara	petin
	Palmer R., Q.	irre	—	—
	Coen R., Q.	tschuru	a	alanne
	Mapoon R., Q.	tschuru	a	aranne
	Gudang, C. York	arta	arta	tetur
Torres Strait	Kowrarega, Torres St.	geta	geta	tara
	Saibai I., N. G.	getö	köigursara	—
New Hebrides	Aulna Malikula	verna	ngorongor verna	tangala ngor. ver.
	Nguna	nangmelearuna	nakinina	natapalakisina
	Aniwa	norima	matshikorima	taperima

Leg	Thigh	Calf	Foot
leoonyana	—	—	luggana
lugguna			lugganah
luggra	tula (J.)	—	lug
langna			dogna
latanama (L.)	kaarwerrar (N.)	warkellar (N.)	pere (E.)
—	dbiran	guram	dyinang
—	karip	kar	dyinong
karip	karip	kar	dyina
—	karip	kar	dyina
—	—	—	—
karnuk	korebuk	tyuluk	dyinañuk
—	karip	pirn	dhinang
—	kureep	beern	dhenung
prum	krip		
kiangi	kiripi	toolangiani	mamberi
	dyeran	born	dyean
karrewa	—	—	jinno
	—	·	jeenong
buyu	darrang	mungo	dinnang
darra (C.)	—	—	menoe (H.)
—	bulloinkoro	wolloma (*turra, pl.*)	—
buiyo	durra	wuruka	dinna
terang	terang	buyu	dhinang
dundu	tburra	—	dinna
matta	dowel	matta	jenna
—	thunda	mullatha	jina
gura	—	—	jina
—	chewen-ta	murnoo	chin-na
taruki	ngulde	kur	turni
wita	kanti	—	idna
mitti, yerko	kanti, mitti	yillamuka	tidna
mungka, yelkko	mungka	yelkkerra	tinna
oora (*pl.*)	thara	thilchaundrie	thidna
mura	muti	kombo	tini
nooldu	nooldu	nooldu	yanar
daonda (P.)	macka (C.)	morna (P.)	queealka (C.)
—	moorn	—	ummal
kalar	tyer	kalar piun	mel
—	—	—	burdro
dangalar	namur	—	nimbilar
—	—	—	nimbal
—	lupara	—	inka
gor (*below knee*)	gudhul	—	tel
malpin	warper	ngar	jinner
—	amathling	—	annil
teni	vwongge	avarri	kwe
teni	wongge	avarri	kwe
utronya	etena	—	oquarra
tirra, ngar	kapi	—	kuku
tete, ngarö	madu	ubalmadu	azuianna, tsanö
neluan	—	—	neluan
natuana	namaona	natorena	natuana
novai	nobili	kagavai	novai

Groups	Dialects	Toe	Tail	Skin
Tasmania	East	—	manna poonee	lurentanena
	South	—	pugghnah	lurarunna
	West and N. W.	—	—	—
	North	rigl (*claw*)	—	kidna
	Miscellaneous	lagurner (N.)	—	loantagarnar (N.)
Victorian Region	Yarra R.	bububidyinang	moibo	dhabo
	Lal Lal	ngatongi dyinong	dorok	mityuk
	Ercildoune	dyina	berkuk	mitch
	Avoca R.	bapdyina	birikuk	mityuk
	Broken R.	—	—	
	Gunbower	dyinañuk	pirikuk	mityuk
	Warrnambool	—	wirang	murnong
	Mortlake	dhenung	weerung	moorn
	Booandik, S. A.	teena	—	moorn
	Lower Lachlan and Murrumbidgee	parthangi	berkoi	looko
	Gippsland	dhakirdyean	wrak	derrauin
	Barwidgee, Upper Murray	jinno	—	wahno
	Woorajery Tribe, Upper Murray, N.S. W.	—	—	—
N. S. Wales and S. Queensland	Wiraidhuri	—	don	iren, yulain
	Turuwul, Port Jackson	dunna (*foot*, R.)	toon (H.)	baggy (H.)
	Awabakal	tinna	—	bukkai
	Kamilroi	—	—	yuli
	Kabi, Mary R., Q.	dhinang	dhun	kubar
	Warrego, R., Q.	darda-dinna	doon	yourring
IV. Australia and West Central	Toodyay (Newcastle)	—	nandi	booka
	Pidong	billbu (*pl.*)	nundi	wanndu
	Minning	—	nuenndi	waiyul
	Lake Amadeus	pulca	whip-poo	bung-kee
South of S. Australia and East Central	Narrinyeri	turnar	kaldari	wankandi
	Parnkalla	—	kadla	piyi
	Adelaide	—	worti	parpa
	Darling	merloo	koondara	pultta
	Diyeri	thidnamookoo (*pl.*)	noora	dula
	Murunuda	tina	kuni	kula
	Mythergoody	yanartinjul	waltha	peer
North Australia and Central Australia	Larrikeya	kwiellgwa	—	beaeeaba (C.)
	Woolna	tyanumunger	—	—
	Daktyerat	ne (*big toe*)	womo	karalla
	Ruby Ck., Kimberley	—	—	—
	Napier Range, Kimberley	—	—	—
	Sunday Island	—	—	—
	Macdonnell Ranges	inkaganya	bara	yinba
	Walsh R., Q.	—	—	—
	Bloomfield Valley, Q.	moroon	piji	youlburn
	Palmer R., Q.	—	—	atteen
	Coen R., Q.	otroo	peanne	kai
	Mapoon R., Q.	otroo	peanne	kago
	Gudang, C. York	dyuro	opo	equora
Torres Strait	Kowrarega, Torres St.	—	koba	purra
	Saibai I., N. G.	—	kupalabö	gungau
New Hebrides	Aulua Malikula	ngorongor neluan	garna	nakolukte
	Nguna	nakini ni natuana	napuena	nawilina
	Aniwa	matshikovai	nosiku	nokeri

BLOOD	BOWELS	EXCREMENT	URINE
warrgata	tiakrangana	tiamena	mungana
coccah	poine	tiannah	munghate munghabeh
—	—	—	—
bolouina	—	—	—
balooyuna (S.)	—	tyaner (N.)	moongbenar (N.)
gurk	dhalandhalarm	gunang	balk
kuruk	dyurung dyurung	—	—
korok	winipa	kunna	kyie
kuruk	burakuk	—	—
—	—	—	—
kurkuk	kulonguk	gunañuk	keñuk
kerek	poraniyung	kunang	keink
goorek	marung	—	—
kro	—	—	—
karku	pilporkeonango	koonangon	keemon
kuruk	kraiuk	kuanang	werak
—	goonoonau	—	—
—	—	—	—
guan	—	dagu	—
banarang (H.)	—	—	—
kummara	konung konaring	konung	keilai
guë	—	—	—
kakke	gunang	gunang	kabur
guing	kurrikurri	—	geeya
noba	gabbil	guner	goonba
yallgu	—	—	—
—	—	—	—
midgee	—	—	—
kruwi	mewi waltyerar	kunar	—
kartintye	kudna	kudna, karta	kumbu
karro	kudna	kudna	kumbo
kondara	koonna-wulkka	koonna	tippara
koomarie	koodnaundrie	koodna	—
katebuca	kalo	wapilinga	pura
gooaroo	oondoo	oondoo	kiperer
dumitilla	namanamanak (C.)	moonmar (C.)	—
mumallweer	—	moonma	—
padawo	wuneru	wuin	wuru
—	—	—	—
—	—	—	—
ilga	—	—	—
alua	—	atna	—
—	—	—	—
yawul	—	—	—
mooler	tchool	tchatcher	yiwan
onyel	—	oothun	—
trellem	loimmi	arri	ambwo
njima	loimmi	arri	ambwo
etunya	ilpi (C.)	onna	ombo
kulka	—	—	—
kulka, kirerö	maita, gabumaita	—	—
menri	mertina	kabakab	meme
natra	namaritana	natae	namemeana
toto	avava	tai	tavaimeme

Groups	Dialects	Food	Live (Verb)	Die
Tasmania	East	—	—	—
	South	—	—	—
	West and N. W.	—	—	—
	North	—	—	—
	Miscellaneous	gibby (N.)	—	mata (E.)
Victorian Region	Yarra R.	dhangitch	murundaka, ngalandf	wegat
	Lal Lal	kutkut	muron	dirta'a
	Ercildoune	tyakol	moronaia	tidaiin
	Avoca R.	dyakitch	muron	titaiang
	Broken R.	—	—	—
	Gudbower	pangguk	murun	wadhyingdha
	Warrnambool	takyir	buindin	kalpurnan
	Mortlake	dhukeanu	—	—
	Booandik, S. A.	—	—	—
	Lower Lachlan and Murrumbidgee	takoori	pooksomaoki	berathin
	Gippsland	lok	—	dhatigan
N. S. Wales and S. Queensland	Barwidgee, Upper Murray	—	—	—
	Woorajery Tribe, Upper Murray	—	—	—
	Wiraidhuri	dangung	murronginga	hallunna
	Turuwul, Port Jackson	dunmingung (R.)	—	—
	Awabakal	—	moron	tettibuliko
	Kamilroi	—	—	baluni
	Kabi, MaryR., Q.	bindba	murubaman	baluman
	Warrego, R., Q.	widgey	kurrin	ballyah
IV. Australia and West Central	Toodyay (Newcastle)	—	—	winnit
	Pidong	—	—	—
	Minuing	—	—	—
	Lake Amadeus	—	—	—
South of S. Australia and East Central	Narrinyeri	takuramb	tumbe	porn il
	Parnkalla	mai, bulta	warrirriti	padhutu
	Adelaide	mai, paru	purruttendi	padlond
	Darling	—	—	bookka
	Diyeri	booka	—	—
	Murunuda	kuti	kuntawanro	balindu
	Mythergoody	putthale	—	—
	Larrikeya	mayoma	amedip (I live)	belingying
North Australia and Central Australia	Woolna	muma	—	—
	Daktyerat	miyu	dukmaadeung	padthadeung
	Ruby Ck., Kimberley	—	—	—
	Napier Range, Kimberley	—	—	kurdiman
	Sunday Island	—	—	eembal
	Macdonnell Ranges	ntutamea	—	iluma
	Walsh R., Q.	jil	—	lon
	Bloomfield Valley, Q.	mie	tchakoi	warli
	Palmer R., Q.	athenning	—	—
	Coen R., Q.	adhou	jeroome	avoinne
	Mapoon R., Q.	adhou	loinre	tomandschooni
	Gudang, C. York	aiye (C.)	—	—
Torres Strait	Kowrarega, Torres St.	—	—	dadeipa
	Saibai I., N. G.	ai	—	umanga
New Hebrides	Aulua Malikula	navangan	timaur (3rd per. sing.)	timis (3rd per. sing.)
	Nguna	navinanga	mauri	mate
	Aniwa	akai	mouri	mate

Eat	Drink	Sleep	Sit
tughlee, tuggana tughrah, tuggmnah —	longholec nugara —	lony roroowa (J.) —	mealpugha crackena (J.) meevenany (J.) —
kible tegurner (N.) tangarabian gutyilin tyakik dyakilan —	kible temokenur ngubian ngupilin kobilang ngupilan —	nenn here logurner (N.) yïmanan kumba komba kumba —	medi (P.) ngalambanan pura nyanga pura —
tyakalang takin dhukeino — yakandin	kupalangga tatthin dhut thunoo tata koopori	kumbandum yuwan uwona looma kombathe	nañok kuppa neenkuka — yantha
dhaando —	dhaando —	berndan —	ñindu —
—	wïjela	—	—
dara patty (C.)	— weede (H.)	ulinga nangara (H.)	guabinga, winga gnallon (H.)
tukkilliko tali dhau, dhoman thenmugga nanang	pitulliko ngarugi dhathin thenmugga gabby nanang	ngarabo babi buandomathi nunamugga bigar	yellawolliko nguddela ñinaman neamugga nit
ngulla, ngurrna warra-maowud null-goonie (pr. p.) takkin melata maiendi tiee tiami tundu —	ngulla, ngurrna — — murttun — — toonjala thapana tundu napa —	nguba kudnaiella — tantin meya wanniti — emargala mookalie parindo wongil	nyinna ningarn oin-nann lewin ikkata tikkandi neengga armuna kunda yïnar
gugai (I eat) —	anjarra (C.) —	allinmingaligal mudgi (C.) va-aqua	aginda (C.) loorl
lakadema niungari	durkadema —	ngurngur adeung munya	adini —
karpe	kuing	kurlin	—
unggarïli al-gooma (W.)	woral lorilama, ñuma	unggerlmo ankuindama	unggalant —
yug ngougal	— boumbi	nog wauni	onjek boundi
athathi angwonogoome ngwonnokomme atedurra purteipa	athathi — tediang wonokomme unggin-ga wanipa	enthul — anronokomme eremadin uteipa	— — angea engka tanureipa
ai pourtanö angkani nganikani kakeina	wanin timin munu keinu	utoi len maturu mero, komero	apatanu ambalok tro natano nofo, konofo

Groups	Dialects	Go	Come	Tell
Tasmania	East	tawe	talpeyawadeno	—
	South	tawkwabee	tutta watta	—
	West and N. W.	tawe	—	—
	North	—	—	—
	Miscellaneous	tagurner (N.)	tecaner (N.)	carne (J.)
Victorian Region	Yarra R.	yananan	birnun	dhumbak
	Lal Lal	yanno	wata	keaka
	Ercildoune	yaanbang	wardiga	ngalayanik
	Avoca R.	yannga	wataga	kiyaka
	Broken R.	—	—	—
	Gunbower	waranggom	yukaiyanok	keap manyup (*tell somebody*)
	Warrnambool	yanan	watta	kaipa
	Mortlake	yanakie	kaka-watakie (kaka, *here*)	leek pukie
	Booandik, S. A.	yanka	kuki	kepa
	Lower Lachlan & Murrumbidgee	yangatbie	tolanden	ngetthelotoona
	Gippsland	yangon	ngauandhi	dhuna
	Barwidgee, Upper Murray	yagamilla	yangabailla	—
N. S. Wales and S. Queensland	Woorajery Tribe, Upper Murray	ya	—	—
	Wiraidhuri	yannana	yawai (*come here*)	—
	Turuwul, Port Jackson	yunda (*go away*, R.)	cowee (*Come on*, H.)	—
	Awabakal	uwolliko	uwolliko	wiyelliko
	Kamilroi	—	taiyanani	—
	Kabi, Mary R., Q.	yanman	baman	yaman
W. Australia and West Central	Warrego, R., Q.	yenmugga	thineyenmugga	thergara
	Toodyay (Newcastle)	watow	yale	nunda wanka
	Pidong	yannma	yannaji	—
	Minning	—	—	—
	Lake Amadeus	by-enie	al-learie	why-talla
South of S. Australia and East Central	Narrinyeri	ngowalour (*imper.*)	arndu (*p. pa.*)	rammin
	Parnkalla	ngammata	budnata	wadlata
	Adelaide	murrendi	kawai	pudlondi
	Darling	—	—	—
	Diyeri	pulkami	kapara	worapami
	Murunuda	cinda	cidinakurna	kawi
	Mythergoody	wobarloo	kowar	minbar
North Australia and Central Australia	Larrikeya	begari (P.)	nallak (C. *come on*)	—
	Woolna	berroque (*go away*)	nallak (*come on*)	—
	Daktyerat	boïadung	baadung	taueradema
	Ruby Ck., Kimberley	—	—	—
	Napier Range, Kimberley	yellar-bonar	yellar-bonar	nigra
	Sunday Island	—	—	—
	Macdonnell Ranges	artyilanama, lama	bityima	albemelama
	Walsh R., Q.	lugar	tok	—
	Bloomfield Valley, Q.	toongi	kutti	milbi
	Palmer R., Q.	—	—	—
	Coen R., Q.	—	—	—
	Mapoon R., Q.	ianganoome	tanoombanne	tschui
	Gudang, C. York	—	impebino wai (C.)	ekalkamurra
Torres Strait	Kowrarega, Torres St.	—	uleipa	mulepa
	Saibai I., N. G.	pa usaru (*go away*)	aie	—
New Hebrides	Aulua Malikula	emben	tipen	bitene
	Nguna	pano	umai, pei	noa
	Aniwa	fano, roro	my	ontuena

SPEAK	WALK	RUN	BRING
pueellakanny	tahlyoonere	rene	kunny wattera
poeerakunnabeh	lawtaboorana	lugara	kunna watta
pooracannaby	—	mella (J.)	—
kane	tagna	tablene pinikta	—
—	tabelti (L.)	noongbenar (N.)	worrar (N.)
durnmln	yaninbulonda	wurwon	tuabongak
geela	yanni	wate	mutyaka
gyigi	yaanbol	plrpa	mutyaka
worake	yannga	piripa	nıutyaka
—	—	—	—
wurake	yanok	wiri	—
lakkan	purpa	wirakan	womba
lukukie	yanakie	karowukie	yanbarnan (*bring that*)
lanka	yanka	wraan	mana
limbi	yena	waiwi	nıanakia
dhuna	—	yangon	wonai
—	biela	pinnela	—
—	—	burrabari	burruma
yarra	yannagagi	bunbanna	bariamalbillinga
byalla (H.)	—	chawa (*imper*. R.)	—
wiyelliko	—	murraliko	—
goalda	—	punagai	taikane
yaman	yanman	bldhallman	barlman
yarra	yenmugga	thungymugga	thinekanımugga
wanka	watow	yatagaly	berrang
wonga	yarra, yanna	bukalli	kunngani
—	ngaraliguni	wudnayeri	—
—	yan-ning	—	—
yamin	ngoppun	kldein	—
wanggata	ngukata	—	kattiti
warrabandi	murrendi	watpandi	kattendi
—	wonga	kolyara	wottolay
yathami	—	—	—
namingi	cinda	namini	kulkini
—	woobar	banjar	buterain
—	akgarni (C.)	muddli (P.)	gurimakerk (P.)
weeyer	mogwerie	moquel	lineter (*bring here*)
lamadung	damadung	tagatadirang	wabagadung
—	urna	—	—
nigra	wongi	mucheri, gurdi -	—
—	ungwarn	tunggora	—
ankama	—	unti (W.)	ngetyima
—	—	anbel	—
koko balkal	toongan toongi	jinbal wari	woondil yara ondo
—	agullaki	—	—
—	—	—	—
tschonokomme	iange	tschiatschine	wia
ekalkamurra	watungi (C.)	—	—
—	—	ringa	—
—	gurgu usaru	—	ngapamani
tisur	penepen	anrui	ti levembene
pasa	surata	sava	piragi
kontucua	katakaro	tere	amyane

Groups	Dialects	Take	Lift	Carry
Tasmania	East	nunne	—	—
	South	nunnabeh	—	—
	West and N. W.	—	—	—
	North	—	—	—
	Miscellaneous	—	—	—
Victorian Region	Yarra R.	kungak	dambok	waronggobok
	Lal Lal	mutyaka	waimok	wamok
	Ercildoune	mutyake	waiwa	mutyaka
	Avoca R.	mutyaka	waiwok	tyulnak
	Broken R.	—	—	—
	Gunbower	manakinyok (?)	waiok	wakura
	Warrnambool	maana	keranga	wombangin
	Mortlake	natonhatnobe ＼	keeraniukie	womburnong
	Booandik, S. A.	mana	—	kinepa
	Lower Lachlan and Murrumbidgee	manapa	wama	karatha
	Gippsland	katbokaia	yenna	kortba
	Barwidgee, Upper Murray	nunda	—	worrongahra
	Woorajery Tribe, Upper Murray, N.S.W.	—	—	—
	Wiraidhuri	barramarra	—	ganna, dummira
	Turuwul, Port Jackson	—	—	—
N. S. Wales and S. Queensland	Awabakal	mantilliko	puromilliko	kurriliko
	Kamilroi	—	tiome	wombailona
	Kabi, Mary R., Q.	komngan	bunma	wombalithin
	Warrego R., Q.	thirrykanga	kundamurra	wodderra
IV. Australia and West Central	Toodyay (Newcastle)	—	yerrup (*lift up*)	berrang
	Pidong	munnma	—	—
	Minning	—	—	—
	Lake Amadeus	mum-jeeli	—	—
South of S. Australia and East Central	Narrinyeri	pultin	preppin	thuppun
	Parnkalla	mankutu	pirriti	milliti
	Adelaide	—	—	nammandi
	Darling	—	—	—
	Diyeri	—	—	wolthami
	Murunuda	cirka	pardunakamana	circinda
	Mythergoody	goorealu	woolbalamar	weejaramar
	Larrikeya	dap, biner	biddbiddla (P. *lift it*)	bonani (P.)
North Australia and Central Australia	Woolna	—	—	—
	Daktyerat	waadema	dalwaadema	karatyadema
	Ruby Ck., Kimberley	—	—	—
	Napier Range, Kimberley	—	—	. —
	Sunday Island	—	—	—
	Macdonnell Ranges	—	tyunama	ngama
	Walsh R., Q.	nig	haratik	nig
	Bloomfield Valley, Q.	woondil	tchara koolpal	woondil
	Palmer R., Q.	—	—	—
	Coen R., Q.	—	—	ajannokomme
	Mapoon R., Q.	proe	angapange	lainre
	Gudang, C. York	—	—	—
Torres Strait	Kowrarega, Torres St.	meipa	—	ang-eipa
	Saibai I., N. G.	—	tiridisa	patauradisö
New Hebrides	Aulua Malikula	tilere	intu	ingunta
	Nguna	trape	trape rakate	trape trape, ova
	Aniwa	amki, amo	tshicitshia	amounga

Make	Break	Strike	Fight
—	—	luggana golumpte	miamengana
—	—	lunghana	moymengana
—	—	—	meniana (J.)
—	—	—	—
pomale (J.)	crackerpucker (N.)	riagurner (N.)	menana (L.)
munggok	kalbornangat	dhilbongalin	dhalgak
wangu	bukomo	dyilpo	bityiring
muyuboga	kalbonga	datyarop	datyarop
muyapok	kalpaiyang	kilpark	tyiltyarang
—	—	—	—
borgok	bukain	dhaka	dhakdyarip
muyubam	mambunga	porta	portapan
koorangong	kindarnong	bardano	burbunallganaka
—	wirlpana (*pp.*)	—	weanban (*pr. p.*)
konga	muruda	tukka	tikaria
ngunauwa	kolakan	koladyin	bondyin
—	—	—	bieba
—	—	paiam	—
bunmarm, marra	—	bumarra	bumallana
—	—	—	—
umulliko	kilpaiyilliko	bunkilliko	wuruwai
baia	gunni	bumale	—
yanggoman	buriman	bunbaman, baiyiman	baiyulaiyu
thenarra	dummerra	bungga	bumbarley
binney	dakkan	boomer	bakash (?)
mulla	kurrdagula	binnya	binnya
—	—	yaguku	paijaji
—	—	poong-an-yee (*pr. p.*)	—
winnin	luwun	mempin	yoyangi
—	kulata	kundata	ya arriti
wappendi	biltilendi	—	—
—	—	pertta	goorinya
—	—	dieami, nundra	thirric
kaivi	tricinda	dilpinda	partindra
pini	tutawar	booer	boonjabinju
godlum (*he has made*)	—	—	berramellidyim
—	moque	—	wauikatpi
dyenadema	taptadema	eadema	murkadema
—	—	—	—
—	—	—	—
—	—	—	—
mbarama	kabuluma	ntyilbutilama	—
—	—	—	—
balkal	tog	donyin	donyin
	toombar	koonil	kooniwe
—	—	—	—
anniingenne	mbwe	tschauogoome	annebe
nringanne	mbwoige	lenyookoome	boipre
—	aterumbanya	untondunya	—
tatureipa	tideipa	matumeipa	—
aimanö	—	urimanö	silamai
mugea	mokot, mambur	atampsea	nabura
mari	marikpori	kpokati	marimari
imna, mo	efatshi	tshi	tatowa

Groups	Dialects	Kill	Fall	See
Tasmania	East	mienemiento	—	mongtone
	South	wanga (D.)	—	nubratone
	West and N. W.	—	—	—
	North	—	—	lamunika (J.)
	Miscellaneous	crackerpucker (N.)	—	neunkenar (N.)
Victorian Region	Yarra R.	tirdowak	baurdangak	nangak
	Lal Lal	ditgundyirin	ba'oorin	ñaalin
	Ercildoune	bandyarang	boika	wari
	Avoca R.	ditguna	boika	ñalalang
	Broken R.	—	—	—
	Gunbower	burngoonin	boikin	ñauwonda
	Warrnambool	—	yungyer	ñake
	Mortlake	balrokonong	yandar burdien	nawhano
	Booandik, S. A.	—	—	naa
	Lower Lachlan and Murrumbidgee	peronmin	poikin	waretemingo
	Gippsland	buladyin	blakgitdualan	dakan
N. S. Wales and S. Queensland	Barwidgee, Upper Murray	—	—	nahga
	Woorajery Tribe, Upper Murray, N.S.W.	binjilgerri	—	—
	Wiraidhuri	ballubundambirra	barguranna	—
	Turuwul, Port Jackson	—	—	gna (C.)
	Awabakal	tettibungngulliko	ngarongaro	nakilliko
	Kamilroi	balubuma	bundane	ngummi
	Kabi, Mary R., Q.	baiyīman	bumbilin	nomngathi
	Warrego R., Q.	bummurra	warra	na-mugga
W. Australia and West Central	Toodyay (Newcastle)	wining	dabat	jenning
	Pidong	—	wannia	—
	Minning	—	—	—
	Lake Amadeus	poong-an-ie	won-enie (*pr. p.*)	—
South of S. Australia and East Central	Narrinyeri	mempin	pingkin	nakkin
	Parnkalla	kundata	worniti	nakkuttu
	Adelaide	padloappendi	womendi	nakkondi
	Darling	bulka	beekka	bommee
	Diyeri	—	poorina	—
	Murunuda	partindra	palinda	kalinatu
	Mythergoody	booanoo	kalganoo	nungarmunu
North Australia and Central Australia	Larrikeya	begilla (P.)	beraddbing (P.)	nagalidja (C.)
	Woolna	—	—	ianungama
	Daktyerat	adinyirkadema	talkadema	dukadema
	Ruby Ck., Kimberley	—	—	—
	Napier Range, Kimberley	—	—	nigra
	Sunday Island	—	—	—
	Macdonnell Ranges	tyakama	iknima	airima
	Walsh R., Q.	—	—	—
	Bloomfield Valley, Q.	yarkin boungal	tari	naimer
	Palmer R., Q.	—	—	tarti
	Coen R., Q.	norpaini	dshoini	tschini
	Mapoon R., Q.	noambwini	dshoini	konatschini
	Gudang, C. York	—	—	ikinya (C.)
Torres Strait	Kowrarega, Torres St.	dadeima	pudeipa	yaweipa
	Saibai I., N. G.	mataman	—	imanö
New Hebrides	Aulua Malikula	tarapee	wiah	lise
	Nguna	maripunue	trowo	punusi
	Aniwa	tshi mati	kotu, koto	koweitia

Hear	Know	Think	Grow
toienook bourack	—	—	myallanga bourack
wayee	tunapee (J.)	—	mangapoiere
—	—	—	mallacka
—	—	—	—
ngarngak	nonondhangyun	nononngarngun	korin korin
ngarwilln	burdu'auan	ngyanin	karingga
yenim	kapang	kapang	karinga
ñanili	dyiangan	ngyanin	karinga
—	—	—	—
ñarnolanda	ñarnolanda	ñarnolanda	koringa
wanga	dheama	—	kareda
wangano	—	—	papgoobun
—	—	ngendon (*pr. p.*)	—
nangon	ngetgathnaining	nangonraorina	krananga
wanggan	kalandanngat	kalandanngat	bernak
—	—	—	—
—	—	—	—
—	winnanggaduringa	winnangganna	yuranna
—	—	—	—
ngurrulliko	—	koteliko	ponikulliko
winungi	tirune	—	—
vrongaman	vrongaman	vrongaman	dhuruman
namiyu	namurriu	binnamebiu	duri
dwanga	kattik	kattik	—
ngunngula	—	—	—
—	—	—	—
co-leenie (*pr. p.*)	—	—	—
kungun	nglelin	kungullun	kringgun
yurranniti	yungkutu	—	mirrurriti
—	tampendi	paiendi	yerthondi
tulleetee	—	moorra	—
—	—	oondrami	boonka
pulo	kalinatu	kalinatu	kurinda
nungun	nunthanu	nunduanjilingu	jnnberingu
baleitong (F.)	alleitong (P. *1st per. s.*)	—	—
tauerema	tauerdyaurera	taueradeung	parkmorema
—	—	nillar	—
—	—	—	—
argutilama	ilbankama	yalama	mankama
—	—	—	—
milgabonimer	ngatchinger	ngonour tchamal	tchire mal
—	—	—	—
namenni	parakooti	—	tsheritte, abvoru
konümenni	parkwiggi	worrokwi	tsheritte, woru
—	—	—	—
krangipa	—	—	—
karnaiginga	mulaigö	—	—
enrongo	enrongobisea	enranea	titarump
trogo	atae	mitrotroa	ulua
fakaragua	keiro	mentua	somo, vere

Groups	Dialects	Give	Like	Marry
Tasmania	East	tyennabeah (*in East or South*)	—	—
	South	—	—	—
	West and N. W.	—	—	—
	North	—	—	—
	Miscellaneous	—	—	—
Victorian Region	Yarra R.	wungak	nininbothombunan	birnibonwarin
	Lal Lal	wa'ak	botyimoan	kurtak
	Ercildoune	wokagan	dhalkuk	mandyarup
	Avoca R.	wukak	wutyapoman	mandirauil
	Broken R.	—	—	—
	Gunbower	wunganda	ñuka	manakiña
	Warrnambool	yungama	—	—
	Mortlake	wookakin	noitcho	wogagae
	Booandik, S. A.	woa, oka (*pr. p.*)	kroamona (*I love*)	manan-woo
	Lower Lachlan and Murrumbidgee	wooki	gnetemowa	—
	Gippsland	yuadhai	magleanman	wandyokan
	Barwidgee, Upper Murray	uga	—	undangyalla
	Woorajery Tribe, Upper Murray	—	—	—
	Wiraidhuri	ngunna	—	—
	Turuwul, Port Jackson	—	—	—
	Awabakal	ngukilliko	—	bumbilliko
	Kamilroi	wune	—	—
	Kabi, Mary R., Q.	woningan	kawun	bindhamathi
	Warrego R., Q.	newa	—	—
	Toodyay (Newcastle)	yunga	mucine	kalla nujet
	Pidong	—	—	—
	Minning	—	—	—
	Lake Amadeus	you-i	—	—
	Narrinyeri	pempin	pornun	napwallin
	Parnkalla	nungkutu	mundalyabmiti	kantyiti
	Adelaide	—	muiyu mangkondi	—
	Darling	—	—	—
	Diyeri	yinkuna (*pr. p*)	—	—
	Murunuda	nonginta	patchi	nyuaringda
	Mythergoody	yumebain	marinjerbuthalbu	narthierejergabu
	Larrikeya	nagok(P. *2nd pers. s.*)	budbaleitmaong (P.)	—
	Woolna	gunmitja (P.)	—	—
	Daktyeral	angadema	elelmok dapadema	—
	Ruby Ck., Kimberley	—	—	—
	Napier Range, Kimberley	younga	—	—
	Sunday Island	—	—	—
	Macdonnell Ranges	ntema	nergama	eknuma
	Walsh R., Q.	nyini	—	—
	Bloomfield Valley, Q.	tchaimer	wahoumal	muniur munmer
	Palmer R., Q.	—	—	—
	Coen R., Q.	naje	njia	nandranne
	Mapoon R., Q.	yia	njia	konoondranne
	Gudang, C. York	utera	—	—
Torres Strait	Kowrarega, Torres St.	pibeipa, wicpa	—	—
	Saibai I., N. G.	paibanö	—	—
New Hebrides	Aulua Malikula	levesak	makapsi	elab
	Nguna	trua	trakiusi	pitauri, laki
	Aniwa	tufwa	hepe	masike tafare mari

SING	WEEP	TIRED	YES
lyenny	naoutagh bourack, tagara toomiack	pryennemkoottiack	narrawallee
lyenne	moi-luggata, tarra toone	kakara wayalee	narra warrah
—	—	—	narro barro
kanewedigda	gnaiele	—	—
carnerwellgurner (N.)	—	—	erre (P.)
yengak	mardun	barnburnguriman	nye
yingile	longga	tirmilin	ye ye
yengarop	yeria	damalang	ye ye
yingile	yirea	burtabaiyang	ye ye
—	—	—	—
naribilip	lumill	mikonda	ngungoi
lirpin	wirpa	wawunga meringa	ku
leit bealun	karartmung	barbumiango	go
nuripa (*impf.*)	loonga	toonking	ngan
yarkoi	looma	mailpalooko	yeai
wadboalan	noön	yardoman	nga nga
gudba	—	—	woorri
—	—	—	aaryama
babbirra	mombanna	birra, gunno	ngawa
bornya (H.)	tonga (H.)	yareba (C.)	mo rem nie (C.)
wittilliko	ngurrunborburrilliko	pirra (*to be*)	—
bao-illona	yugila	malo gini	yo
duppathin	dunginan	ngaiya balun	yauni
youngey	wongey	gilyapairliyon	kairla
wanga baket	dup	winkin	kwa
warrilla	ngola	thallthinnya	ngow, kun
—	milellinug-nginn	—	oh, nadenn
war·rannie (*pr. p.*)	ho-lan-yee (*pr. p.*)	tarn-tun-perrin	—
ringbalin	parpin	nguldamulun	katyil ng-ng
kuri kundata	ngattutu	innelli	nga, ya, yandi
palti mutandi	murkandi	mentamentanendi	ne, tiati
yengko	neerra	—	marrayta
wonka	yindrami	patbuna	kookoo, kow
eilcinda	youcinda	nocipinda	youi
piala	paringu	lergingu	ner
gugumal	billum	annelling (P.)	koo, goo (C.)
meninyer (*pr. p.*)	—	inmokiter	gogo
nanama adini	werkmadini	digarap (*sick*)	ya
—	—	—	—
—	—	wirigeo (*sore*) -	ku
—	—	—	—
ilima	itnima	aranta, borka	wa
—	oggui	winyi	—
bouri doudal	bati	bujerbouran	yea
—	—	—	yowo
ndmnagoome	tae	arrauenyumenne	nge
anjanyakomme	pfie	arraunrumenne	ya
—	—	—	ia
sagul piyepa	—	—	wa
—	maindi	—	wa
engake	antang	nerambauta kaskas	e, e
lenga	kai trangi	mawosa	iora, io
feke tagora	tagi, kotagi	taru	keini, bo

Groups	Dialects	No	I	Mine
Tasmania	East	parmgarah, noia	—	meena
	South	timeh, pothyack	meena, manga (J.)	-mea (*suffix*)
	West and N.W.	mallya leah	—	—
	North	—	—	—
	Miscellaneous	nendi (P.)	mana (P.)	
Victorian Region	Yarra R.	yuta	marambik	marambaiak
	Lal Lal	boraka	bangangik	bangurdidyik
	Ercildoune	ngalanya	wangal	wangin
	Avoca R.	ngalanya	wan	warnguk
	Broken R.	—	—	
	Gunbower	barápa, brapa	ngai, ngatch	yikek
	Warrnambool	ngi'ngi	ngatuk	ngatunat
	Mortlake	bangadong	mathuk	athongmet
	Booandik, S. A.	ngi-ing	ngatho	ngananine
	Lower Lachlan and Murrumbidgee	warti	ynethi	naika
	Gippsland	ngalgu	ngaiyu	ngitalung
N. S. Wales and S. Queensland	Barwidgee, Upper Murray	oneugaba, baal	—	—
	Woorajery Tribe, Upper Murray, N.S.W.	woori	athoo	—
	Wiraidhuri	wirai, barre	ngaddu	—
	Turuwul, Port Jackson	beall (C.)	gnia (C.)	dannai (C.)
	Awabakal	keawai	ngatoa	emmongta
	Kamilroi	kamil	ngaia	ngai
	Kabi, Mary R., Q.	kabi, wa waka	ngai, adhu	nganyonggai
	Warrego R., Q.	walla	nunthey	—
W. Australia and West Central	Toodyay (Newcastle)	wadder	—	—
	Pidong	waddji	ngutha	ngunnathung
	Minning	yanngun	—	nunnga
	Lake Amadeus	we-umpa	niyouloo	—
South of S. Australia and East Central	Narrinyeri	nowaiy ng·ng	ngape	nganauwe
	Parnkalla	madla, kutta	ngai, ngatto	ngaitye, ngaityidne
	Adelaide	madlanna	ngai, ngaityo	ngaityunna
	Darling	nahtta	ahppa	—
	Diyeri	abi	althoo	nie
	Murunuda	waba	ungaro	ungaro
	Mythergoody	umbi	nigo	nigeringu
North Australia and Central Australia	Larrikeya	alika (C.)	ana ananga	anege
	Woolna	leita	tanunga	unggoingee
	Daktyerai	aka	nga	nga ngave
	Ruby Ck., Kimberley	—	—	—
	Napier Range, Kimberley	marla	ni	—
	Sunday Island	—	—	—
	Macdonnell Ranges	itya	yinga, ta	nukara
	Walsh R., Q.	—	du (?)	—
	Bloomfield Valley, Q.	kari	aio	aiko
	Palmer R., Q.	anuncha	inun	—
	Coen R., Q.	njianni	yupoo	tanoome
	Mapoon R., Q.	njee	iange	tamre
	Gudung, C. York	untamo	uba (C.)	—
Torres Strait	Kowrarega, Torres St.	longa, guire	ngai, ngatu	ngow (*m.*) udzu (*f.*)
	Saibai I., N. G.	maigi, launga	ngai	ngau
New Hebrides	Aulua Malikula	a, o	anu	tuknu
	Nguna	e	kinau	aginau
	Aniwa	jimra	avou	tshaku

Me	Thou	Thine	Thee
mina	—	—	—
meenah	neeto	-eena (*suffix*)	neeto
—	—	—	—
pawahi (P.)	morambina	morambaiak	—
bangik	bangin	—	—
—	dalkukar	dalkukwangin	—
wangin	war	—	—
ngyikin	ngindi	ngindi.	ngindi
ngatuk	ngutuk	ngutunat	ngutuk
meindook	—	—	—
ngatho	ngooro	nganaon	—
ynethi	ynyaa	—	—
ngidha	—	nginalunga	—
athoo (?)	enoo	—	—
ngannal	ngindoo	—	—
emmong	bi	—	bin
ngunna	nginda	nginnu	nginnuna
nganna	ngin, ngindu	nginyonggni	nginna
nunthey	—	—	—
nanye	—	—	—
—	yinnda	yinndong	—
niena	yentoo	—	—
ngan, an	nginte	ngumauwe	nguni
ngai	ninna, nuro	nunko, nurko	ninna
ngai	ninna, ninko	ninkunna	ninna
—	indoo, imba	—	—
anie	yondru	—	ninna
ungaro	—	—	—
anannga (P.)	ityenna	ityennege	—
unggoingee	—	—	—
erin	nun	nungbe	nundyu
yingana	lenkina, nga	unkwanga	unkwangana
du	—	—	—
enya	youndo	youno	yina
tano	andramme	angenoome	ngonoo
tano	andreamme (?)	angeoomre (?)	ngeanoo
nna	ngi, ngidu	yinu	—
—	ngi, ngido	—	—
nnu	engko	takengko	engko
au	nigo	anigo	ko
avou	akoi	tshow	akoi

Groups	Dialects	He	His	Him
Tasmania	*East*	—	—	—
	South	nara (J.)	—	—
	West and N. W.	—	—	—
	North	—	—	—
	Miscellaneous	narrar (N.)	—	—
Victorian Region	*Yarra R.*	kannuk	kathup	—
	Lal Lal	giawa	wanyuk	—
	Ercildoune	—	—	—
	Avoca R.	kinyuwa	wanyuk	—
	Broken R.	—	—	—
	Gunbower	maalu	maikatch	maalu
	Warrnambool	ngulampe	—	ngulampe
	Mortlake	mange yananee	—	—
	Booandik, S. A.	nung	noongerengat	—
	Lower Lachlan & Murrumbidgee	—	nooka	kikinga
N. S. Wales and S. Queensland	*Gippsland*	—	—	—
	Barwidgee, Upper Murray	—	—	—
	Woorajery Tribe, Upper Murray, N.S.W.	—	—	—
	Wiraidhuri	nilla	—	—
	Turuwul, Port Jackson	—	darringal (C.)	—
	Awabakal	—	—	—
	Kamilroi	ngernia	ngerngu	—
	Kabi, Mary R., Q.	ngunda	ngundano	ngunda
	Warrego R., Q.	numbu	numbuka	numbuka
W. Australia and West Central	*Toodyay (Newcastle)*	—	—	—
	Pidong	ball	ballong	—
	Minning	—	—	—
	Lake Amadeus	—	—	—
South of S. Australia and East Central	*Narrinyeri*	kitye	kinauwe	kin, ityan
	Parnkalla	panna, padlo	parnuntyuru	panna
	Adelaide	parnu, parnuko	parnukunna	parnu
	Darling	wahtta, wahto	—	—
	Diyeri	nouliea	noonkanie	nooloo
	Murunuda	—	—	—
	Mythergoody	—	—	—
North Australia and Central Australia	*Larrikeya*	bienneba	biennege	yaba (P.)
	Woolna	—	owingee	owingee
	Daktyerat	yundun	yundunde	ne
	Ruby Ck., Kimberley	—	—	—
	Napier Range, Kimberley	—	—	—
	Sunday Island	—	—	—
	Macdonnell Ranges	era	ekura	ekurara
	Walsh R., Q.	—	—	—
	Bloomfield Valley, Q.	nulu	ngongo	ngongonin
	Palmer R., Q.	—	—	—
	Coen R., Q.	lopoo	ngonoome	ngorpe
	Mapoon R., Q.	leo	niamroo	ngoa
	Gudang, C. York	—	—	—
Torres Strait	*Kowrarega, Torres St.*	nudu, nue	nunue	nudu, nue
	Saibai I., N. G.	ngoi	ngungu	—
New Hebrides	*Aulua Malikula*	hena	tahena	hena
	Nguna	nae	aneana	a, e, sa, nia
	Aniwa	aia	tshana	aia

We	Ours	Us	You
—	—	—	neena
—	—	—	neena, nee
—	—	—	—
—	—	—	—
warrander (N.)	—	—	ninga (D.)
morombolok	moroinbongata	moroinbongata	moromnguta
moroninyala (*you and I*)	—	—	—
bangitok	wangitok	wuringiting	—
dbalkukangal	wanginurak	—	—
wangu	wangitok	wuinanding	—
—	—	—	—
yangur	yangurau	yanguren	—
—	—	—	—
pulijah	—	—	—
ngathoe, ngathoat	ngana-anu	—	ngootpaler
youngoun	—	—	—
—	—	—	—
—	—	—	—
—	—	—	—
—	—	—	ngindugir
—	—	—	ngeene (C.)
ngeanni	—	—	—
ngearun	—	—	bula (*dual*)
ngeane	ngeanengu	—	ngindai
ngalin	ngalinngur	ngalin	ngulani
nunna	nurraka	nunna	yindu
nundo, nanye	—	nunda, nanye	nunda
ngulli (*dual*)	ngullingu	—	—
—	—	—	—
—	—	—	—
yentoo-nully (*dual*)	—	—	—
ngurn, ngele (*dual*)	ngurnauwe	nam	ngune, lom (*dual*)
ngarrinyelbo	ngarrinyelburu	ngarrinyelbo	nuralli
ngadlu	ngadlukunna	ngadlu	na, naako
—	—	—	—
uldra	jannanie	iana, alie	yoora, yinie
—	—	—	—
nulyindu	unarar	—	yundu
dorendera	dorennege	—	gugurangura
—	—	unggoingee	neelana
auur, ergur	auure, ergure	erpuro, erguro	nungur
—	—	—	—
`—	—	—	jok
—	—	—	—
anuna	anunaka	—	arankara, nibala
de	—	—	—
angin, ali (2) ana (3)	anginunger	anginin	yourer, youbal (2)
—	—	—	inoo
boitti	namboome	boumbwoonime	andrappu
mboi	nianrume	mbwonoonie	andreu
—	—	—	unduba (C.)
arri, albei (*dual*)	arrien, albcine (*dual*)	—	ngitana, ngipel (*dual*)
—	ngabanu	—	ngitamura, ngipel
antil	tahantil	antil	amuntil
ningita	aningita	ngita	nimu
ncitin	tshote	acitia	acowa

GROUPS	DIALECTS	YOURS	YOU (OBJECT)	THEY
Tasmania	East	—	—	—
	South	—	—	nara (J.)
	West and N. W.	—	—	—
	North	—	—	—
	Miscellaneous	—	—	—
Victorian Region	Yarra R.	nguta	—	konoit
	Lal Lal	—	—	—
	Ercildouue	—	—	—
	Avoca R.	—	—	—
	Broken R.	—	—	—
	Gunbower	ngudhek	ngudhek	—
	Warrnambool	—	—	—
	Mortlake	—	—	—
	Booandik, S. A.	ngootpalerorong	—	nungpaler
	Lower Lachlan and Murrumbidgee	—	—	—
	Gippsland	—	—	—
N. S. Wales and S. Queensland	Barwidgee, Upper Murray	—	—	—
	Woorajery Tribe, Upper Murray, N.S.W.	—	—	—
	Wiraidhuri	—	—	ngannaingulia
	Turuwul, Port Jackson	ngeenede (C.)	—	—
	Awabakal	—	—	bara
	Kamitroi	—	—	ngarma
	Kabi, Mary R.,Q.	ngulamo	ngulambola	dhinabu
	Warrego R., Q.	yin-ga	—	thenna
W. Australia and West Central	Toodyay (Newcastle)	—	—	—
	Pidong	—	—	—
	Minning	—	—	—
	Lake Amadeus	—	—	—
South of S. Australia and East Central	Narrinyeri	nomauwe	ngune	kar, keengk (*dual*)
	Parnkalla	nuralluru	nuralli	yardna
	Adelaide	naakunna	naa	parna, parnako
	Darling	—	—	—
	Diyeri	yinkanie	—	thana
	Murunuda	—	—	—
	Mythergoody	—	—	goonulnoorloo
North Australia and Central Australia	Larrikeya	gurennege	—	bedenbera
	Woolna	netangee	—	—
	Daktyerat	nunguro	nunguro	wurundun
	Ruby Ck., Kimberley	—	—	—
	Napier Range, Kimberley	—	—	—
	Sunday Island	—	—	—
	Macdonnell Ranges	aragankara	—	etna, eratera (*dual*)
	Walsh R., Q.	—	—	—
	Bloomfield Valley, Q.	yourunger	yourunin	tanner, buller (a)
	Palmer R., Q.	—	—	—
	Coen R., Q.	yamboome	neappi	lorpe
	Mapoon R., Q.	ranrumme	neanne	lorpi
	Gudang, C. York	—	—	inyaba
Torres Strait	Kowrarega, Torres St.	ngitanaman	—	tana, pale (*dual*)
	Saibai I., N. G.	ngalpan	—	tana
New Hebrides	Aulua Malikula	tahamuntul	amuntul	bera
	Nguna	animu	mu	nara
	Aniwa	tshowa	acown	acre

THEIRS	THEM	YESTERDAY	TO-DAY
—	—	nentegga menyena	—
—	nara (J.)	neea nunnawa	—
—	—	—	—
—	—	—	—
moroinbatbana	thanan	mulongmulok	walden-pont (J.)
—	—	taliyo	yilnbo
—	—	dhyalige	iniriyo
—	—	talige	nyawiu
—	—	—	nauwiyo
—	—	kyilikyilik	kilauitch
—	—	ngaangat	tigape
—	—	akatho	makateba
nungpalerat	—	—	keto
—	—	kilonaki	kilmaki
—	—	bukang	dhilai
—	—	—	—
—	—	—	—
—	—	ngingurain, gambai	ngidyigallila
—	—	boorana (H.)	yagoona (C.)
burunba	barun, bulun (dual)	—	buggai
—	—	gimiandi	ilanu
dhinabuno	dhinabubola	ngamba	dhall, gilumba
therraka	yellowdirry	gunda	kainyi
—	—	kuochat bennang	yaye
—	—	—	—
—	—	—	—
—	—	—	—
kandauwe	kan	watangrau	hikkai nunggi
yardnakkuru	yardna	wiltyarra	yattanyarru
parnakunna	parna	—	yellara
—	—	—	—
thananie	thaniya	illahgo	keilppo
—	—	urukuli	—
—	—	genodljodl	kuri
bedennege	—	goolawa (C.)	ilangua
—	—	winemegwa	targenail
wurundunde	wuru	pendyodin	aman
—	—	—	—
—	—	—	miliar (now)
—	—	—	—
etnika	—	tonurka	lata
—	—	—	—
tannunger	tannunin	yili	yenenya
—	—	—	—
—	—	anunba	amilmean
neroomyunoome	neru	angoinne	oragokoo
lornrumme	lorne	agwoinye	kaidakke
—	—	yulpu (C.)	ura (C.)
tananian	—	ngul	—
tanamunu	—	wargaiga	kaiba
tahera	hera	nino	abakal
areara	ra	nanova	masoso
tshare	acre	neinafe	iranei

Groups	Dialects	To-morrow	Where are the Blacks
Tasmania	*East*	—	—
	South	—	—
	West and N.W.	—	—
	North	—	—
	Miscellaneous	—	—
Victorian Region	*Yarra R.*	buiburuing	windya yang golin
	Lal Lal	yiranmu	wiya koli
	Ercildoune	barpobarp	windyala krutang
	Avoca R.	barapa	windya koli
	Broken R.	—	
	Gunbower	pirpu	windyalo yuanuk kuli
	Warrnambool	tunggat	windana maarban
	Mortlake	malungiba	wondha mara
	Booandik, S. A.	kalapa	winthowoongi tonoro
	Lower Lachlan and Murrumbidgee	koongonda	—
	Gippsland	brundu	wunman konai
N. S. Wales and S. Queensland	*Barwidgee, Upper Murray*	—	—
	Woorajery Tribe, Upper Murray, N.S.W.	—	—
	Wiraidhuri	—	—
	Turuwul, Port Jackson	parrybuga (H.)	ware (*where*, H.)
	Awabakal	kumba	—
	Kamilroi	nguruko	tulla murri
	Kabi, Mary R., Q.	yirkī	weno dhan
	Warrego R., Q.	burda	deam bulla maiing
W. Australia and West Central	*Toodyay (New-castle)*	bennang	yungar wingal
	Pidong	—	—
	Minning	—	—
	Lake Amadeus	moong-al-yer-roo	—
South of S. Australia and East Central	*Narrinyeri*	ngrekkald	yangi narrinyeri
	Parnkalla	malturlo	—
	Adelaide	paninggolo	—
	Darling	wahmbeenya	weendyah wimbaja
	Diyeri	—	—
	Murunuda	—	—
North Australia and Central Australia	*Mythergoody*	wargumurra	wundoo narjerar
	Larrikeya	emangua (C.)	arabelidjee belira (C.)
	Woolna	melarnga	looarkieinga ungalooqua
	Daktyerat	nungoyune	ngan yao ngaran
	Ruby Ck., Kimberley	mukamukan	—
	Napier Range, Kimberley	mini warar (*by-and-by*)	jenar wamba
	Sunday Island	—	—
	Macdonnell Ranges	ingunta	—
	Walsh R., Q.	—	—
	Bloomfield Valley, Q.	woongoon	bummer wonjarin
	Palmer R., Q.	oloong	—
	Coen R., Q.	woingatimmi	—
	Mapoon R., Q.	pronganne	nambarra andrangoo
	Gudang, C. York	achunya	ama undukera (C.)
Torres Strait	*Kowrarega, Torres St.*	batteingh	—
	Saibai I., N. G.	bangal	—
New Hebrides	*Aulua Malikula*	mebko	asamangk miet aranembi
	Nguna	matamai	natang moli loa manga wai
	Aniwa	aratou	togata pouri wehe

I Don't Know	How	Who	What
—	—	—	telingha tebya
—	—	—	pallawaleh
—	—	—	tarraginna
—	—	—	—
—	—	—	wanarana (P.)
windhongga	kurndirnar	kunup	winnar
wlya	ñuran	wela	winyar
ngalanyanga	ñangura	windyaro	windya
windya	ñonguran	winyaro	ñangomin
—	—	—	—
windya	ñongurarau	winyar	ñañuk
ngurtambu wirn	windhigunga	ngara	ngana
wangatong	—	—	nana
—	—	nganoo	nan
warthenete naagana	—	nenga	nungoa
ngolangat boangan	—	ngan	ngan mandyi
—	—	—	—
—	—	—	—
—	—	—	minyang
manyero (H.)	—	—	—
—	—	gan	—
—	—	andi	minya
wa vronga	minanggo	ngando, ngangai	miñanggai
nai-ma	dirraga	narnna	minyan
nanye katti wadder	—	—	nait
yurilo	thann	nganna	na
—	—	—	—
—	—	—	—
nowaiy ap nglemin	mengye	nganggi	minyi
mintlali	wantye	nauwe, nganna	nauwe
—	ngaintya	ngendo, nganna	ngaintya, nganna
yoongahnjy	—	—	minna
—	wodow	warana	mina
unugunarar	unginju	urnu	uni
elabauna (C.)	—	harbira	analla (P.)
illebidbanna	—	—	—
anungi dyauermagiere	anungenung	angun	nigida
—	—	—	—
—	—	—	—
—	—	—	—
yuka	—	nguna	iwana
jer	—	—	—
aio natchimul	wonjere	wonjoungou	wanu
—	—	—	—
—	—	andrakoo	annai
—	—	aye	annai
che (C.)	—	—	—
—	—	ngadu, nga	eimi
ngai karawaigo	—	nga	miai, mida
ien, anu selisembosen	mobah	hase	nepah
a ta atae a mau	trapale sava	sei	nasava
avou puspusi	kontucua	akai	taha

Groups	Dialects	When	Where	Why
Tasmania	East	—		—
	South	wabbara	ungamlea	—
	West and N. W.	—	—	—
	North	—	—	—
	Miscellaneous	—		—
Victorian Region	Yarra R.	mulugo	windya	winyirangga
	Lal Lal	willang	wīya	wekarok
	Ercildoune	pirbanyuin	windyalar	windyaii
	Avoca R.	ngirtoge	wīndya	ñangur
	Broken R.	—	—	—
	Gunbower	naturuk	windyalo	naturuk
	Warrnambool	windagadha	winda, windagara	ngangaranuk
	Mortlake	—	woonaha	
	Booandik, S. A.	nawet	na	nukine-waa
	Lower Lachlan and Murrumbidgee	wutti	narrakanian	nungora
	Gippsland	nara	wunman	nannane
N. S. Wales and S. Queensland	Barwidgee, Upper Murray	—	—	—
	Woorajery Tribe, Upper Murray, N.S.W.	—	—	—
	Wiraidhuri	widyunga	—	wargu
	Turuwul, Port Jackson	—	warè (H.) wutta (R.)	—
	Awabakal	ba	—	—
	Kamilroi	wīru	tulla	minyago
	Kabi, Mary R., Q.	weño	weñomini	miñani
	Warrego R., Q.	wonding	thirring	minyangor
West Central	Toodyay (Newcastle)	—	winga	nagook
	Pidong	nunnga	thulla	—
	Minning	—	—	—
	Lake Amadeus	—	—	—
South of S. Australia and East Central	Narrinyeri	yaral	yangi	mengye,
	Parnkalla	warpara	watha, wana	ngannaru
	Adelaide	nallaalatti	wa, wada	—
	Darling	—	weendya	minna
	Diyeri	wintha	wadarie	minandroo
	Murunuda	—	—	—
	Mythergoody	ungeebura	wondu	wondare
North Australia and Central Australia	Larrikeya	—	harguaarabelidje (C.)	arbiddla (P.)
	Woolna	—	ungalooqua	—
	Daktyerat	anyikading	ngaran	ngandukmane
	Ruby Ck., Kimberley	—	—	—
	Napier Range, Kimberley	—	—	—
	Sunday Island	—	—	—
	Macdonnell Ranges	—	ntana	woka
	Walsh R., Q.	—	—	—
	Bloomfield Valley, Q.	wunjere wunjere	wondenya	wanuringo
	Palmer R., Q.	—	—	—
	Coen R., Q.	andraimenni	andrenne	anaiki
	Mapoon R., Q.	andraume	andrangoo	anaikatti
	Gudang, C. York	—	undukera (C.)	—
Torres Strait	Kowrarega, Torres St.	—	anaga	mīpa
	Saibai I., N.G.	—	nago	—
New Hebrides	Aulua Malikula	nengesa	ambe, nembe	mebah
	Nguna	seve rangi	seve tokora, wai	ekasana
	Aniwa	inaia	webe	tiaha

One	Two	Three	Four
marrawah	pia wah	lia winnawah	pagunta
marrawah	pooalih	talleh	wullyawa
—	—		—
pammere	kateboueve		—
marai (P.)	boula (J.)	wyanderwar (N.)	
kanbo	bondyira	bindyir ba kanmerng	bindyir ba bindyira
kuimat	buletch	buletpainnot	boletch ba boletch
kaiyap	polaitch	polaitch bo kaiap	polaitch bo polaitch
kaiap	buletch	buletchkaiap	boletch ba boletch
—	—		
kaiap	buledya	buledya kaiap	bulet bulet
kaiapa	bulaitcha	polinmea	bulaitcha bulaitcha
kiapa	bulitha	puligmea	puligmea
wando	boolite	boolite ba wando	boolite ba boolite
yetina	polatol	polatol yata	—
kutupon	bulumon	bulumon kutuk	bulumon bulumon
—	—	—	—
oonbi	bulla	bulla oonbi	bulla bulla
ngunbai	bula	bula ngunbai	bunga
wogul (C.)	boola (C.)	brewy (C.)	—
—	buloara	—	—
mal	bular	guliba	bular bular
kalim, kualim	bulla	bulla kalim	bulla bulla
yoummun	kubbo	kubbolana youm	kubbolana kub
kain	guchal	mow	—
kutia	kutharra	murrngul	—
kaiaddnu	kutharra	warrul	—
gooch-a-goora	go-darra	mun-kuripa	—
yamma-laityi	ninkaiengk	neppaldar	kukkuk
kubmanna	kuttara, kalbelli	kuppo, kulbarri	—
kuma	parlaitye	marukutye	yerrabula
neecha	boolla	bollaneecha	boolla-boolla
koornoo	mundroo, bolya	parkoola	mundro-la mundro-la
ururu	pagoli	pagoli ururu	—
pigundul	gurtho	gurtho gurion	umbigal
kalaguk	galatilik	galatilik kalaguk	galatilik galatilik
tillingita	toloya	toloya thidle	toloya ma toloya
yaunuka	verenuka	wirittyauen	verunverun
yangga	kujara	tilowaji	—
wingair	kujara	kujara lina	kujara kujara
aringk	kwîr	iridhar	kwîra kwîr
ninta	tera	teramininta	teramatera
gatim	bul	artu	alpun (*many*)
nupoon	marmara	koloor	kakouar
appool	impa	aroolko	abunji
pemi	ambodbu	tshumajum	—
pemi	adbuti	tshumayum	—
epiamana	elabalu	dama	—
warapune	quassur	uquassur war	uquassur uq
urapon	ukasara	uka mondobigal	ukauka
bokol	enrua	entil	embis
sikai	trua	trolu	pati
tase	erua	toru	la

Groups	Dialects	Five	Ten
Tasmania	East	pugganna	—
	South	marah	—
	West and N. W.	—	—
	North	karde	karde karde
	Miscellaneous	—	—
Victorian Region	Yarra R.	bindyiro ba bindyiro kanbo	wurtona
	Lal Lal	boletch ba boletch ba koi-motch	bolen mirna
	Ercildoune	kaiya manga	bolaimanya
	Avoca R.	boletch ba boletch ba kaiap	boletch manya
	Broken R.	—	—
	Gunbower	bulet bulet kaiap	—
	Warrnambool	—	bulatya ba bulatya
	Mortlake	puligmea	pulig mara
	Booandik, S. A.	—	—
	Lower Lachlan & Murrumbidgee	ninumanyi	kinoneto murnangi
	Gippsland	yailmon (?)	—
N. S. Wales and S. Queensland	Barwidgee, Upper Murray	—	—
	Woorajery Tribe, Upper Murray, N.S.W.	bulla bulla oonbi	mutto (*many*)
	Wiraidhuri	—	—
	Turuwul, Port Jackson	—	—
	Awabakal	—	—
	Kamilroi	mulanbu	bulariu murra
	Kabi, Mary R., Q.	murin	—
	Warrego R., Q.	kubbolana k. y.	—
IV. Australia and West Central	Toodyay (Newcastle)	—	—
	Pidong	—	—
	Minning	—	—
	Lake Amadeus	—	—
South of S. Australia and East Central	Narrinyeri	kuk kuk ki, keyakki	—
	Parnkalla	—	—
	Adelaide	—	—
	Darling	—	—
	Diyeri	mundroo-mundroo-koornoo	—
	Murunuda	—	—
	Mythergoody	—	—
	Larrikeya	kuiare	binolka
North Australia and Central Australia	Woolna	—	—
	Daktyerat	yangaramotung	mundul (*plenty*)
	Ruby Ck., Kimberley	—	—
	Napier Range, Kimberley	—	—
	Sunday Island	ara ara	alburi
	Macdonnell Ranges	—	—
	Walsh R., Q.	—	—
	Bloomfield Valley, Q.	warpool	warpool (*many*)
	Palmer R., Q.	—	—
	Coen R., Q.	—	—
	Mapoon R., Q.	—	—
	Gudang, C. York	—	—
Torres Strait	Kowrarega, Torres St.	uq. uq. warapune	—
	Saibai I., N.G.	ukaukamodobai	—
New Hebrides	Aulua Malikula	elima	sangabul
	Nguna	lima	rualima
	Aniwa	erima	tagahuru

B

L

T

Y

Z

Printed by BALLANTYNE, HANSON & Co.
London & Edinburgh